Negotiating Hollywood

Negotiating Hollywood

The Cultural Politics of Actors' Labor

Danae Clark

University of Minnesota Press
Minneapolis
London

A different version of chapter 5 appeared as "Acting in Hollywood's Best Interest," *Journal of Film and Video* 42.2 (1990): 3-19. Used by permission.

Published by the University of Minnesota Press
111 Third Avenue South, Suite 290, Minneapolis, MN 55401-2520
Printed in the United States of America on acid-free paper

Library of Congress Cataloging-in-Publication Data

Clark, Danae.
    Negotiating Hollywood : the cultural politics of actors' labor / Danae Clark.
        p.   cm.
    Includes bibliographical references and index.
    ISBN 0-8166-2544-1
    ISBN 0-8166-2545-X (pbk.)
    1. Motion picture industry—Economic aspects—United States.
    2. Trade unions—Motion picture industry—United States—History.
    3. Industrial relations—United States.   4. Motion picture industry
    in motion pictures.   I. Title.
    PN1993.5.U6C54   1995
    384'.83'0973—dc20   94-46538

In memory of my father

Contents

Preface

This book is not about stars. Nor is it a book that could easily be categorized as belonging to that area of film scholarship typically called star studies. It is rather a book that seeks to destabilize the concept of "star" and to interrogate the very premises upon which "star studies" traditionally rests.

Having initially set out to write a book about stardom in the U.S. film industry, I was somewhat surprised by this position of disavowal. But, in retrospect, this position is a logical—or, at least, not too surprising—outcome of my attempts to confront and rearticulate the shifting relations of theoretical practice within film studies.

In the mid–1980s, when my ideas for this project were just beginning to take form, the field of film studies was in a peculiarly intense period of flux. *Screen* theory, which had played such a dominant role in film studies over the previous decade, was losing its theoretical potency. Cultural studies, in its imported form from Britain, was meanwhile gaining theoretical and political ground. These two approaches to the study of film (or popular media more generally) seemed to lock heads.[1] The former was overtly textual and psychoanalytic, taking into account our subjective and unconscious relations to film and filmic (dis)pleasure. The latter was more contextual, theoretically and materially rooted in the lived practices of cultural struggle. Where did star studies fit within this tension?

Star studies, by this time, was an accepted and established subgenre within film studies. In the wake of Richard Dyer's influential book *Stars* (1979),[2] analyses of stars or star images combined semiotic and poststructuralist theory with subcultural and sociological approaches in ways that were beginning to challenge the text-based criticism of *Screen* theory and to incorporate aspects of cultural studies. It was in part this refashioning of the field and the hybrid approach to the study of stars that drew my intellectual interest. But by the late 1980s it appeared that the field of star studies was spinning its wheels. Most articles about stars followed a standard formula: choose a well-known star, establish his or her star image by analyzing his or her roles in various films or the "persona" established for the star through studio publicity departments, then explore the possible effects that this star image might have on spectators. When inflected by psychoanalytic theory, these analyses additionally stressed the way in which the star figured as an "ego ideal" for the spectator and was representative of or resistant to the standard terms of sexual difference.

A few scholars challenged or were working outside of this formulaic frame. Barry King and Richard deCordova, for example, were concerned with the star system as an economic and enunciative apparatus within the film industry (see chapter 1).[3] Others, such as James Naremore, became interested in the aesthetics of performance.[4] But, in my mind, the initial excitement surrounding star studies as a potential arena for challenging current modes of thought within film scholarship had been lost. There were boundaries that needed to be pushed, and I set out to find them.

Given the ascendancy and prominence of cultural studies within the academy, it seemed particularly odd that the study of stars had not moved further in this direction. Work that embraced a cultural studies perspective, which includes the proliferation of articles and conference papers on Madonna during this period, exploited the subcultural aspect of cultural analysis but did little to challenge the Dyerian notion of the star image as a polysemic sign to be read and interpreted. Thus, while cultural studies was being criticized elsewhere for its emphasis on reception and its methods of determining (subcultural) reading practices, star studies seemed to look the other way.

Armed with my own suspicions about reception studies, as well as with my background in Marxism and political economy, I began to resituate the star within the sphere of production. I had been inspired by Janet Staiger's analysis of labor unions and their relation to Hollywood modes of production in Bordwell, Staiger, and Thompson's *The Classical Hollywood Cinema*.[5] Her description of the political and social economies of film production pried open an aspect of film history that was rarely talked about (and never taught) within film studies. Yet her contribution to this volume did not easily mesh with the neoformalist analyses of film texts offered up by the rest of the book, and her treatment of labor suffered from an objectivism that characterizes much historical writing. Further research on the star system and actors' labor within the U.S. film industry yielded similar results. What did exist—and there wasn't much—was devoid of theoretical reflection. Film history appeared to be a victim of cultural lag; in the frenzy of textual analysis during the 1980s, historical work was put on a back burner. And though revisionist film histories enjoyed a phase of popularity during this period, their revisionism did not bring together theory and history in ways that complemented the critical force of contemporary theoretical discourses.[6]

One of the main attractions of cultural studies is that it provides a way to contextualize history within a theoretical frame. It does not simply present the events of times past as historical fact, but sees history as a terrain of power upon which various institutions, individuals, and groups struggle to constitute social practices. These practices, although articulated in ways that are concrete, are always contingent and continue to be embroiled in relations of power that struggle to define them historically or to appropriate their historical use for current social practices. Cultural studies is furthermore motivated by a political agenda that focuses on and privileges the voices and struggles of social subjects who historically have been silenced within the institutionalized relations of dominance under capitalism.

I realized that pushing star studies to the furthest extent of this cultural studies frame would require a thorough rearticulation of the field. First and foremost, the cultural studies emphasis on the struggles encountered by social subjects (or the relations of power that constitute subjects economically, politically, and discursively) meant that stars must be endowed with subjectivity. This move was in direct opposition to standard star studies analyses, which posited the star (image) as an object/text/sign to be read and interpreted by the spectator. To regard the star as a social subject who struggles within the film industry's sphere of productive practices would shatter the spectator's fetishistic relation to the star as an ego ideal and break the spectator-as-subject versus star-as-object binarism that implicitly formed the base of star studies as we know it. Much of my initial project, then, was a matter of thinking through the terms by which one can speak about subjectivity in relation to stars.

The second problem that arose in my attempt to reorient star studies within the theoretical and historical trappings of cultural studies concerned the role of stars themselves. Given the cultural studies emphasis on the social relations of power, it seemed necessary to think of these relations in terms of the entire star system. This meant that my analysis should embrace the conditions and struggles of actors at all levels of the star system hierarchy: some were stars, but most were not. It meant that, in addition to developing a theory of *actors'* subjectivity, I would need to address the social relations among actors and the way in which their various antagonisms and affiliations contributed to an understanding of their subject identities and positionalities.

The third major issue I confronted in this venture was that of history. The cultural studies (re)appropriation of history is rooted in the "conjunctural," a methodological commitment that allows a movement from a general level of abstraction to a concrete locale of specificity. Such a practice does not seek to explain a historical moment in its entirety or to make conclusive claims about the nature of its future configurations. Rather it seeks to understand how certain relations of power enter into the economic, political, and discursive dimensions of cultural struggle within a given historical moment. History, according to a cultural studies perspective, is a hegemonic field in which relations of power are structured in dominance; and social subjects reproduce, struggle against, or attempt to transform existing relational practices. In this process, the historical subject is produced.

This book, then, is motivated by several interlocking considerations. In developing the field of star studies in relation to cultural studies, it is important that many of the basic premises of star studies scholarship be examined and rearticulated to include a space for the actor as social subject and to intervene in the historical formation of this subjectivity. In the first chapter I discuss some of the theoretical barriers that have restricted the development of star studies in this direction and prevented the field from addressing these issues. An ideological complicity with capitalist relations of power, for example, has stunted a thorough examination of the star system and caused scholars to focus their attention only on stars as opposed to workers further down in the labor hierarchy. A theoretical emphasis on film aesthetics and film reception has furthermore led scholars to focus on the image and the spectator-image relationship and to ignore the conditions of labor that produce the image.

Chapter 1 is also devoted to articulating the terms of actors' subjectivity. Through the process of examining and critiquing psychoanalytic and post-Marxist models of the subject, I come to the conclusion that actors' subjectivity must be rooted in a notion of labor power differences. Defined as social subjects in relation to their occupation and position as workers in the film industry, the terms of actors' subject identity are formed materially and discursively through the antagonistic relations of labor practices. The theoretical basis for this concept of labor power differences derives from Marx's theory of labor power, but is rearticulated through post-Marxist theories of the subject that view subjectivity as a contingent and fragmented social construct. I also discuss ways in which labor can be treated as a discourse such that labor power is dislocated from its material, wage-labor roots in the economic sphere, and can be seen to enter discursively into all aspects of the production-exchange process.

Chapter 2 engages the problematic of actors' subjectivity at a general level of abstraction. It examines the way in which the Hollywood star system has structured labor power differences among actors, and between actors and producers, as a way to position actors' labor and subjectivity. It also looks at actors' contracts—the material and discursive sites where actors became wedged between the forces of production (labor) and the forces of exchange (film image)—to determine the means by which studios used them to reinforce social relations of power. The second part of this chapter examines the terms of subject formation (or modes of subjectivity) that conditioned actors' labor prior to the studio era. Here I raise the possibility of the "collective subject" and address the way in which various fragmented notions of subject identity cohered in relation to labor ranking (high versus low) or institutional medium (film versus theater) to form labor unions.

The next three chapters attempt to ground my analysis of actors' labor and subjectivity within a specific historical conjuncture. I have chosen to restrict my analysis primarily to the year 1933—a time of economic and political crisis in the U.S. film industry, and a time during which various discourses of labor were called into conflict. Many of these conflicts were precipitated by the establishment of Franklin Roosevelt's National Recovery Administration (NRA), which guaranteed labor the right to bargain collectively through representatives of their own choosing. Chapter 3 focuses on the attempt by screen actors to achieve self-representation through the formation of the Screen Actors Guild. It provides a detailed historical account of the institutional barriers to unionization, the role the federal government played in mediating or exacerbating the conflicts between actors and studio management, and the difficulties actors encountered in organizing and defining their subjective identities in relation to labor discourses.

Chapter 4 examines the way in which Hollywood studios attempted to undercut actors' self-representation by rewriting labor discourses of the NRA period into discourses of entertainment and morality. It looks, for example, at the way the Motion Pictures Producers and Distributors Association positioned actors as immoral subjects in relation to the goals of the NRA by foregrounding the detrimental effects of star salaries on film production and by highlighting the organization's own commitment to the Film Production Code. It also looks at the role of studio publicity departments and fan magazines in constructing a positionality for actors that drew audience attention away from the goals of labor, reinforced studio discourses of entertainment, and thus sought to deny actors' subjectivity by defining them solely in relation to a salable image.

In chapter 5 I undertake analyses of NRA publicity and two feature-length films produced during this period—*42nd Street* and *Morning Glory*—to determine how their discourses of actors' labor further reinforced studio labor policies. Unlike most textual analyses of recent years, these are concerned not with the ways that spectators may have interpreted these films or were positioned as social subjects in relation to them, but with the ways that the films' labor discourses sought to position actors' labor and subjectivity in the field of commodity exchange. These textual documents are further analyzed in relation to the material practices of labor that were in place at the time of the films' production to indicate the gap that existed between studio discourses of labor and the actual conditions of labor experienced by actors who struggled as social subjects within their confines.

Finally, in chapter 6, I explore the ways that a theory of labor power differences and actors' subjectivity may be applicable to star studies and to the field of cultural studies more generally. I argue that much is to be gained by both fields through incorporating a theory of labor at the heart of their analytical and historical projects. The concept of labor, for example, can potentially revise our understanding of spectatorial practices and the social relations (of power) between spectators and actors while breaking down the production-consumption binarism that underlies many of our current approaches to film study. Before concluding, I also provide some brief analyses of contemporary actors' labor to indicate ways in which scholarship in this area might continue to prosper.

This genealogical narrative is offered as a guide through the terrain of actors' labor and subjectivity—because, through my attempt to resituate star studies within the theoretical, historical, and political project of cultural studies, the terrain may appear unfamiliar to some. Those coming to this work from the perspective of cultural studies (or even star studies) might find the historical detail overwhelming. And historians interested in the details of actors' labor during the studio era might find the theoretical framework too intrusive or cumbersome. But I offer this work, and its accompanying guide, as an opportunity to rethink our relationship to stardom in ways that challenge and destabilize our own positionings as academic subjects.

Acknowledgments

Like many books this one has a long and sometimes tortuous history. So I would like to thank the people who have assisted along the way. My appreciation goes first to the University of Pittsburgh for a Faculty Travel Grant that allowed me to conduct research in Los Angeles in the summer of 1990, and for a research leave in the fall of 1992, which allowed me the pleasure of completing this project while tending to my newly arrived daughter.

A special thanks goes to Mark Locher of the Screen Actors Guild, who gave me special permission to peruse the contents of the organization's library, and to the SAG staff for their generous cooperation in photocopying mounds of requests. I would also like to thank Ned Comstock and Leith Adams in the Warner Bros. Collection at the University of Southern California's Doheni Library; the reference staff in Special Collections at the University of California, Los Angeles, Library; and Howard Prouty and the staff at the Academy of Motion Pictures Arts and Sciences Margaret Herrick Library. Much earlier, but no less appreciated, assistance came from the Interlibrary Loan staff at Miami University who unflinchingly and competently followed through on requests for even the most obscure materials.

Other people who deserve special thanks are Thomas Schatz, for encouraging me to develop this project in the first place; Eileen Meehan, for guiding me through my first encounters with Marxist political economy; Rick Altman, for his intellectual and advisory support; Jane Gaines and Dana Polan, for reading earlier versions of this book and providing feedback and encouragement; and Allen Larson, for assisting in the final stages of the manuscript's preparation. I would also like to thank Janaki Bakhle and Robert Mosimann at the University of Minnesota Press for their editorial support.

My greatest appreciation, however, goes to my partner, Kristina Straub, who has supported me throughout this project and who, quite simply, makes my life happy . . . and to my daughter, Bailey, my greatest source of joy.

1 / The Actor's (Absent) Role in Film Studies

The field of star studies is permeated by discourses of lack. The most common lament speaks to the sheer paucity of work in the field prior to the mid-1970s and the attendant reluctance that film scholars traditionally have shown in allowing the screen actor entrance into the discipline of film studies proper. The lack of "serious" work, in other words, is attributed to an institutional bias that has sought to maintain a distance between the academy and popular experience and to distinguish the intellectual labor of film scholars from what was perceived as an uncritical idolization of stars by nonintellectuals. Thus, even though film scholars may have lacked the appropriate tools for generating substantive work on stars, there was little motivation for developing them.

Richard Dyer's book *Stars*, published in 1979, was pivotal in this regard, showing how semiotics, (post)structuralist, and ideological models of contemporary film studies might be mobilized in theorizing stardom.[1] Generally credited with marking a shift and legitimating the field of star studies, Dyer's text provided a semiotic and sociological framework of investigation that fueled star studies analyses for the following decade. At about the same time, due primarily to the influence of British cultural studies, the field of film and media studies increasingly began to embrace popular culture as a site of analysis. This disciplinary shift helped to remove the taint that once accompanied the subject of stars in academic circles and facilitated the trend toward the critical analysis of stars (though one could argue equally well that the new star studies were themselves a factor in generating scholarly interest in popular culture).

One reason the topic of stars became so compelling is that "the star" provided a nexus for examining relations among the traditionally divergent concerns of industry structure, narrative representation, and viewer identification. By insisting upon "the mode of existence, circulation, and functioning of star images as multi-media, intertextual productions,"[2] star studies began to challenge the text centeredness of previous film scholarship, directing our attention to the way that star images work across genre, beyond narrative, and within different discursive venues (e.g., fan magazines) to establish a variety of identificatory positions for spectators. In addition, film scholars began to find ways to analyze how different modes of acting and performative elements, such as gestures and expressions, contribute to a film's meaning and "frame" the cinematic spectacle.[3] Together, these trends have been filling the "lack" in star studies, enriching our understanding of the construction of star images, the relation between star image and character, and the relation between star image and spectator.

Scholars of star studies, however, continue to point to certain lacks in the field. In the introduction to his anthology *Star Texts*, for example, Jeremy Butler argues that mainstream film theories still elide consideration of performance in favor of film technique and narrative structure—a condition that has been influenced by such prominent theorists as Christian Metz.[4] In *Picture Personalities*, Richard deCordova attends to another lack. "Many studies of the phenomenon of stardom," he says, "situate their analyses at the level of the individual star" while ignoring the role that the star system plays in structuring and producing the individual star image.[5] DeCordova thus turns his attention to the actor's position in the enunciative apparatus of the star system (a move that ironically returns him, via Emile Benveniste, to Metz's theories of film language). He examines how a star might be produced or positioned as a subject within the institutional practices of cinema: "It is not enough to say that the actor appears as subject in film; one must go on to describe the specific conditions of its appearance in discourse, the practices that give rise to this figure and determine its function."[6]

Barry King's work, by contrast, notes the relative lack of emphasis placed on the actor's economic position within the star system. Arguing that "writers on stardom are seemingly obsessed with matters of signification,"

King is concerned with the way the star system "develops out of and sustains capitalist relations of production and consumption."[7] In his attempt to reconcile a political economy approach to stardom with a semiotics of acting, he views the star as a laborer as well as a performance commodity or sign; stardom is as much a process of agency as it is a determined effect of the industry's economic system of managing and differentiating the acting profession.[8]

In addressing the star system as a historically dynamic process of economic and discursive production, the works of deCordova and King have provided some productive alternatives to the second generation Dyerian analyses of stars that dominated the 1980s. Each one's strength, however, appears to be the other's weakness. Although deCordova is concerned with the actor's subjectivity, he fails to endow his subject with agency. Trapped within an enunciative apparatus, stars appear to be effects of the star system, individual paroles to the star system's langue. King, on the other hand, insists on actors' agency but does not ground this agency in a theory of the subject. Like other Marxist analyses, his tends to portray laborers as fully formed individuals who respond, however actively, to established structures of economic and discursive domination.

What is still lacking in star studies, I would argue, is a theory of stardom that builds upon King's and deCordova's contributions by rearticulating the theoretical and historical relations between actors' labor and subjectivity. What is lacking in more specific terms is a *cultural studies* approach to star studies that would introduce the notion of struggle into the scene of actors' labor, necessitating not only an investigation of the individual and collective political conflicts that actors have encountered within the Hollywood production system, but the fragmented and fought-over position of the actor as a subject of film labor and film representation. The purpose of this book, then, is to establish a cultural studies framework that will allow for an exploration of these issues. But before discussing just what this cultural studies approach to stardom would entail, it is important to understand the ways in which the field of star studies has developed thus far. For, as I shall argue, the "lacks" that star studies scholars have produced arise out of the ideological biases that haunt historical and theoretical inquiry in mainstream film studies.

Actors and Historical Discourse

According to Vincent Mosco, "The questions posed and the methods employed [by historians] are, more often than not, guided by the demands of the media industries and the political authorities charged with overseeing their operation."[9] Indeed, it seems that when it comes to discussing the phenomenon of stardom, film scholars have taken their cues from the film industry itself. As deCordova notes, the star system became possible due to "a strict regulation of the type of knowledge produced about the actor," which increasingly emphasized the actors' screen images and deemphasized their behind-the-scene working conditions.[10] The majority of star studies have subsequently engaged in this same sort of knowledge production. Just as the studios diverted audience attention away from the "factory" aspect of the dream factory in order to create more profitable modes of viewer identification with the cinema, film scholars have focused on star images and the production and consumption of "dreams" involved in the star–text relation. By adopting this institutional point of view, film scholars reinscribe the inequities that exist behind the image by refusing to examine the material conditions out of which this signification is produced.[11]

Such tendencies are reinforced by a long tradition of writing film history from the point of view of capitalist interests. (In this sense, actors have not been singled out; the same treatment is reserved for set designers, electricians, carpenters, musicians, and even writers and directors.) This top-down model presents the history of the film industry in terms of linear development, capital growth, and corporate relations while ignoring or undermining the relations and conflicts between labor and management in the production system. From a capitalist perspective, stars are a means (a commodity) to an end (the reproduction of capital). This distinction, which separates human labor from the accumulation of capital, allows for a historical account of the star system that ignores the relations of production. By defining the use value of stars in terms of their exchange value for industry leaders, actors' labor is separated from the process of production and made invisible within the history of industrial expansion.

This explains why so many textbooks of American film history focus only on the *origins* of the star system. Once the use value of the star system is firmly established, there is little need to examine the ongoing and conflicting practices that determine the economic and discursive aspects of stardom. Even in more fully developed histories, the concerns of actors themselves are not addressed. Gorham Kindem, for example, provides a history of stardom that defines the star system as "a widely practiced strategy for securing and protecting production investments, differentiating movie products, and for ensuring some measure of box office success."[12] Richard Maltby elucidates the concerns of management somewhat differently by suggesting that stars played a significant role in perpetuating the ideological construction of Hollywood films "*by offering themselves* to their audiences as idols of consumption" (emphasis added).[13] But because such studies do not address the possibility of actors' agency and resistance—either individually or as part of organized labor—they risk leaving the impression that actors were readily complicit with their employers' desires.

Studies that focus on the careers or images of individual stars comply with the traditional top-down model of history in another way. Simply put, these studies focus on stars. Though the term "star system" refers to the institutional hierarchy established to regulate and control the employment and use of *all* actors, stars have become a privileged class within the division of actors' labor. The character actors and screen extras, who occupy the lower strata of the hierarchy and comprise the majority of the acting profession, receive less attention than their luminous colleagues. A practical difficulty arises, of course, in investigating the historical circumstances of these actors, especially the extras, whose employment was sporadic and largely undocumented. Precisely because screen extras represent the nameless rank-and-file labor force that the capitalist system demands, film researchers are left with no real database from which to proceed.

To date, only a few scholars (most of them trained as historians) have taken on this challenge. The most detailed study of actors' labor prior to 1940 is Murray Ross's *Stars and Strikes* (1941)—a book that, despite its title, delivers an impressive historical account of the labor struggles affecting actors at all levels of the acting profession.[14] David F. Prindles's more recent book, *The Politics of Glamour* (1988), focuses more on the post–World War II era.[15] Other studies—such as Larry Ceplair and Steven Englund's *The Inquisition in Hollywood* (1979) and, to a lesser extent, Michael Nielsen's "Toward a Workers' History of the U.S. Film Industry" (1983)—situate actors' working conditions within the context of other collective bargaining groups in Hollywood.[16] Together these texts redress the imbalance of corporate-inflected histories by employing "bottom-up" modes of explication that make the actor as laborer visible as an agent of history. But while all of these studies provide useful historical information, usually from a labor perspective, none of them provides a theoretically informed approach to labor or to the writing of history. Rooted in conventional standards of objectivity, these historical accounts do not address the issue of subjectivity, which, I will argue later, is a crucial component in theorizing labor from a cultural studies perspective.

There are a few historical studies of individual stars that make connections to studio working conditions or labor–management relations. In an article on Bette Davis, for example, Thomas Schatz notes that "the trajectory of Davis's career . . . is revealing not only of Warners' operations and the evolution of its house style during the 1930s, but also of the contract player's role in the studio production system."[17] In her impressive book *Contested Culture* (see especially chapter 5: "Reading Star Contracts"), Jane Gaines provides numerous examples of how the commodified images of Hollywood celebrities become embroiled in legal struggles and issues of ownership.[18] Most studies on individual stars, however, do not engage with the historical aspects of labor in any direct manner. More interested in the popular construction of star images, or the effect of these constructions on spectators, these investigations emphasize the textual and fetishistic elements of stardom.

Actors and the Aesthetic Tradition

It is impossible to divorce aesthetic considerations from economic ones since aesthetic models or theories of film have traditionally been deployed in the shadow and service of capitalism. Perhaps the best example of this aesthetic/economic complicity is auteur criticism. Although now widely critiqued as a model for film analysis, auteurism still flourishes within star studies by providing the implicit basis upon which their analyses rest. Theorists have attempted to sidestep the problems of intentionality and coherent individualism generally associated with this form of analysis by separating the star (auteur) from the star image (polysemic signifier of meaning) and concentrating their analysis on the latter. But this move does not avoid the political underpinnings of the auteurist approach. Like traditional auteurism, whose task has been to separate out the true creators from the *metteurs en scene*, star studies combine with a capitalist perspective by structurally precluding any consideration of Hollywood's rank-and-file labor force. Scholarly interest is thus limited to those individual stars who, through their natural talent or determination, have "risen above" or transcended the ranks of ordinary craft workers (who fulfill the capitalist demand for efficiency) to produce something of artistic merit. In the case of star studies' appropriation of auteurism, artistic achievement is measured in terms of a star's charismatic and creative ability in constructing a complex polysemic image.

The auteur *system* meanwhile allowed studio heads to differentiate their product while maintaining greater control over labor–management relations. Higher production budgets and greater control over a film's creative process often were used as rewards for good behavior or as lures for higher prestige and better pay. Encouraged to identify themselves as "artists" instead of "workers," auteurs secured their positions by aligning themselves with the interests of management rather than by offering their assistance to those of lower rank who stood to benefit from the auteur's higher position and potential leverage with studio heads. Certain individuals were thus separated from and elevated above their fellow workers as a way to reduce organized labor dissension and to forestall labor organizing while reinforcing the studios' economic and ideological control over artistic expression. As I will show in chapter 3, the screen actors' struggle to unionize in the 1930s was by and large a struggle to overcome the auteurist logic of the star system.

It is worth noting that, although several designated "auteurs" worked during the prestudio era (e.g., D. W. Griffith and Charlie Chaplin), most of the auteur pantheon consists of directors and actors who produced work after 1930, a period of advanced capitalism during which the studios' corporatized hegemony developed greater control over the star system and assembly-line production. Although this period of economic and political domination achieved a form of mechanized efficiency heralded by Western capitalists, its machinelike production ethic threatened to strip its own product (i.e., film) of artistic value. Auteur criticism can thus be read as an attempt to restore the virtues of art to a system of productivity designed to minimize creative difference and individual accomplishment. It is an indication, as Walter Benjamin might have argued, of the way we cling to an aura of art and the cult of personality in an age of mechanical reproduction.[19]

Star studies are further complicated by the aesthetic tradition's fixation on the image. Since actors appear in their commodity form as textual representations, evidence of an actor's work seems to be recorded on the screen in terms of performance, thus causing a confusion between dramatic expression and human labor. This performance is, in turn, inextricably linked to character, narrative, and publicity images. Because of such linkages, film scholars have often conflated the star with the star image or with filmic representation. (This is precisely what happens in "auteurist"-inflected star studies even though they claim to separate the auteur from the image.) The roots of this conflation extend back to the very beginnings of film theory, when Hugo Munsterberg argues that film art depends on "the substitution of the actor's picture for the actor himself."[20] Decades later André Bazin would approach the issue slightly differently by explaining that the actor maintains a paradoxical relation to the image: while "the drama on the screen can exist without actors," actors also become "bound up with the very essence of the *mise-en-scène*."[21] It is the legacy of these early film aestheticians that continues to inform contemporary formulations of screen acting in which actors are alternately equated with the image and superseded by the

image. In "An Aesthetic Defense of the Star System in Films," Maurice Yacowar states the problematic succinctly: "The film actor is all image [but] the
image is more real than the physical entity."[22]

If the image makes the embodied, material actor nonexistent, then the
actor's ability to control his or her own performance is placed into question.
This is the sort of logic that has led theorists of performance to conclude that
screen acting is a submissive form of craft, especially when compared to stage
acting. A screen actor's submission, in other words, is seen as a natural outcome of screen performance whereby the actor becomes an actor only by
yielding to character, to narrative, and to the film's direction. According to
Yacowar, what is "filmic" about screen acting is not so much acting as "pas-
siving," whereas "acting for the stage is projection, something active."[23]
Other critics grant the actor a greater degree of control. James Naremore, for
example, argues that it is a misconception to believe that screen actors are not
acting; even naturalistic methods involve "mastery, skill [and] inventive-
ness."[24] Charles Affron furthermore insists that "the film medium conspires
to free the actor, to relieve him of the burdens of naturalistic portrayal."[25]

But regardless of whether the actor is thought to be in control of his or
her performance, film scholars working within the aesthetic tradition have
not pushed the issue of the actor's control beyond the level of the image. In
two recent texts on actors' performance—James Naremore's *Acting in the
Cinema* and Carole Zucker's anthology *Making Visible the Invisible*—screen
acting is never once referred to as a form of work.[26] Through their priori-
tization of the image, actors' "labor" becomes aestheticized as "expression."
Film scholars "screen out" an economy of work that determines the very
structure of the acting profession. Actors are meanwhile forced to submit to
the image, where they are subjected to scrutiny and objectified in fantasy.

The objectification of stars has served as a source of great pleasure for
film critics (qua spectators). According to Affron, this pleasure occurs because acting "includes the viewer in its mystery . . . and we are active in the
assimilation of gesture, tone, expression, decor, and the general structure that
contains the specific performance we witness."[27] This form of assimilation
leads Laleen Jayamanne to call actors "incarnations of emotions."[28] Actors
may be locked within the surface of the image, she says, but they reveal a
broad range of expression that challenges us to enter the complexities of human emotions. The challenge to read an actor's performance is consequently
taken as an invitation to spectatorial manipulation. Affron provides a rather
blatant example of this in his discussion of Greta Garbo's films:

> Garbo can die for me around the clock. I can stay her [*sic*] in that final moment of her life; I can turn off the sound and watch, turn off the picture and listen, work myriad transformations in speed and brilliance, and then restore the original without losing a particle of its integrity.[29]

The desire to manipulate star images grows out of the pleasure derived from fetishizing stars as objects. As Naremore puts it, the impenetrable barrier of the screen "promotes a fetishistic dynamic in the spectator; the actor is manifestly *there* in the image, but *not there* in the room, 'present' in [an] intimate way . . . but also impervious and inaccessible."[30]

This fetishization of stars and the equation of stars with star images remains the prevailing tendency of star studies. Even Richard Dyer's early work on stars must be understood in relation to this aesthetic tradition. First of all, his emphasis on stars as images in media texts tends to ignore the role of actors' labor in the production process. In addition, given his concern with the significance that star images have for spectators, Dyer promotes a concomitant fetishization of stars and their performances. To be sure, Dyer complicates the Bazinian tradition of examining the revelatory function that the image has for viewers by emphasizing the *ideological* significance(s) of the star image. In *Stars*, Dyer views the star as a "structured polysemy" that embodies a number of cultural contradictions. According to how these contradictions are negotiated, the star image can reinforce or resist "dominant ideology." Thus, "the star's image is characterised by attempts to negotiate, reconcile or mask the difference between the elements, or else simply hold them in tension."[31] The spectator, meanwhile, chooses to identify with, or to manipulate, various aspects of the star image that arise out of star discourse. In his case study of Jane Fonda, for example, Dyer examines ways in which the (pre-1980) Fonda image allowed spectators to identify with Fonda the all-American girl, Fonda the sex symbol, or—for those looking for a resistant reading—Fonda the political radical. He adds that even those star images that appear to reinforce the status quo (e.g., Sandra Dee) can nonetheless open up subversive possibilities for the spectator.

Criticisms of Dyer's work have centered not on his aesthetization of stars but on his notions of star charisma and the spectator's oppositional readings of star images. Pam Cook, for example, suggests that the problems inherent in Dyer's *Stars* stem from collapsing the incompatible projects of sociology and semiotics. "Dyer's use of the sociological approach and his formulations of the notions of 'charisma' and 'identification,' " she says, "run the risk of suggesting that there is no need to change the dominant forms, that *in them-selves* they provide the possibility of the expression of alternative or opposi-tional ideologies."[32] Barry King attacks the issue from the other side, asking, "How is it possible, if one situates the meaning of stardom at the level of the content of audience response, to offer a general theory of stardom?"[33] Perhaps, as Dyer argues, "the audience is also part of the making of the image."[34]

But then, perhaps the subject of analysis is no longer the star.

The "Object" of Study

At a British Film Institute (BFI) workshop on stars in 1982, Christine Gled-hill identified two major issues emerging from the papers and debates: "One concerns where the ultimate site, or sites, of the star's effectivity actually lie; another, the different levels of social reality on which cultural analysis may focus."[35] Since the weekend workshop was conceived of as a forum for ex-tending the points made in Pam Cook's critique of Dyer's work, the partici-pants examined stars from sociological and semiotic perspectives as well as from other perspectives—most notably psychoanalytic theory. The star was alternately defined as "an industrial marketing device," "an artefact of tex-tual production," and "an institutionally sanctioned fetish." But regardless of how the star was defined, the star's effectivity was always examined in rela-tion to people or forces outside itself (i.e., the star had no effectivity in de-termining anything for herself or himself). Whether the issue of stardom was approached from the angle of texts, extratextual discourse, the "cinematic machine," cultural and subcultural values, or unconscious processes, the ul-timate site of the star's significance was located in the viewer.

This workshop was an important step forward in that theories of sub-jectivity played a central role in star studies for the first time. But ironically, the papers that came out of this workshop centered issues of subjectivity on the spectator, not on the actor. This made it clear that, within film studies, theories of subjectivity are essentially theories of *spectatorship*. As BFI partici-pant Anne Friedberg put it, "The star's body is *not* the subject's."[36] Thus, while the actor's body—or, more specifically, the actor's labor—allows for the production of images, the spectator's relation to or identification with these images is what is thought to begin the process of subjectivity in the cinema. One result of this one-sided investigation is that theories of subjec-tivity remain on the consumption-exchange side of the cinematic process and fail to acknowledge the "subjects of production." Like the top-down models of film history and the textual emphasis of aesthetic criticism, theories of spectatorship have thus participated in the scholarly oppression (re-pression?) of film workers, rendering their struggles as political and social subjects invisible.

This omission has created problems in theorizing the actor's conflicting positions or roles within the cinematic institution. Perhaps the major prob-lem of current theoretical approaches is that they trap actors in a "star image" versus "real person" binarism, thus leaving the impression that a fully formed, preconstituted subject exists somewhere behind the image, provid-ing a stable signified for the image signifier. Some theorists have tried to break out of this binarism by proposing the "persona" as a third category that exists somewhere between the real historical person of the actor and the parts played by that actor in individual films,[37] but this triple articulation (person, persona, parts) remains inadequate. Imagine positing an opposition between a spectator's moments of image identification and the "real person" or per-sonality outside the movie theater. Even more absurd would be categorical distinctions among the person, the spectator, and various roles played in the office or home.

If we are to articulate an adequate, or at least a more consistent, theory of subjectivity in film studies, the privileged position of the spectator can no longer force actors to submit to structures that deny the actor's subject iden-tity or that create an "actor as object" versus "spectator as subject" binarism. The term "spectator," which signifies the subject in a viewing relation to screen images, must be joined by the term "actor" to signify the subject in a producing/performing relation to those images. In other words, actors are not that different from spectators. Both actors and spectators must be viewed as heterogeneous subjects who are caught up in a continual process of cul-tural resistance, pleasure, and negotiation. The major difference between spectators and actors, as subjects, lies in their differing relation to the specific social practice of the cinema.

Perhaps not surprisingly, current approaches to film subjectivity (i.e., spectatorship) do not provide automatic insight into the construction of ac-tors' subject identities. The psychoanalytic model, which finds its explana-tion of the split and fragmented subject in an oedipal drama that depends on structures of looking, transfers easily to theories of spectatorship since spec-tators look at and fetishize images on the screen. But how is the actor to be theorized in relation to the scopophilic drive? What does the actor "look" at? What is the object of the actor's gaze? While answers to these questions could undoubtedly be found, they are not crucial to the issue of actors' sub-jectivity. The psychoanalytic emphasis on sexual difference(s) might additionally provide insight into an analysis of actors' employment conditions, film assignments, image constructions, and so on. But sexual difference cannot singularly account for the condition of being an actor. Likewise, to theorize the actor's primary subject formation in terms of an oedipal trajectory (e.g., the mirror phase, castration anxiety), would tell us little about the actor's specific relation to and position within the social institution of cinema.

The major problem that one encounters, then, in applying psychoanalytic theories of spectatorship to a theory of actors' subjectivity is that the former depend upon the actor's role as a fetish object and "ego ideal" for the spectator. Actors remain a passive object of the gaze, and their subjectivity is denied in becoming a fetishized image or pleasurable display for the spectator. The *active* spectator, a more recent arrival on the theoretical scene, can furthermore manipulate the actor/image according to her or his own fantasies. As Cook observes, it is a "small wonder, then, that stars fascinate men and women and that we are reluctant to give them up."[38] But we must be willing to relinquish our fetishistic (and often erotic) relation to actors *and* to psychoanalytic theory if we are to address the actor as an active subject and grant the actor any power or agency in defining her or his own subject identity.

One way to understand the actor as a historical subject who actively negotiates the economic, political, and ideological discourses of identity, is to examine the actor's working relation to the practice of cinema. The notion of labor is crucial here, since employment by the cinematic institution is a primary condition for being a screen actor (just as securing access to the image—through ticket purchase or video rental—is a primary condition for being a spectator). Factors such as race, age, talent, ethnicity, beauty, and sexuality enter into a theory of actors' subjectivity as some of the many discourses that circulate around the figure of the working actor. But the "actor as worker" is a prerequisite term in formulating the actor's subject identity. More than a third term or conduit between person and image (like the term "persona"), the "actor as worker" becomes the site of intersecting discourses involving the sale of one's labor power to the cinematic institution, the negotiation of that power in terms of work performance and image construction, and the embodiment of one's image (on-screen and off-screen) as it becomes picked up and circulated in filmic and extrafilmic discourse.

The "actor as worker" should not be construed as the *true* identity of the actor, but rather as an effective discursive construct that allows a particular political interpretation or ideological understanding of the actor's subject formation. This point cannot be stressed enough since it is meant as a corrective to certain implicit formulations of subjectivity in Marxism. Classical Marxism, for example, situates all social subjects in relation to a class structure determined by the economic base. Subjects are defined quintessentially in terms of class—with preference for the "working class"—and every antagonism experienced by subjects is reducible to a class antagonism. Post-Marxists, particularly the British cultural theorists, have attempted to avoid the problem of class reductionism by arguing that classes "do not have fixed, ascribed or unitary world views."[39] Stuart Hall has noted, in addition, that "the weakness of the earlier Marxist positions lay . . . in their inability to explain the role of the 'free consent' of the governed to the leadership of the governing classes under capitalism."[40] Thus, through the Gramscian concept of hegemony, cultural studies theorists have sought an explanation of power that reworks the relation between classes and meaning systems in terms of struggles over discursive practice.

Still, whether "class" is treated as an indeterminate discursive construct or as a verifiable material category, it tends to remain a privileged factor in (post-)Marxist analysis. As Terry Lovell argues, "Any theory of the individual subject which purports to complement historical materialism must be broadly compatible with Marx's account of the class subject."[41] This imperative, however, poses substantial difficulties for a theory of actors' subjectivity. Class identity not only fails to capture the complexity of an actor's position within the Hollywood production system, it is unclear how—and in relation to whom—that class identity would be formulated. Would one address class divisions between actors and studio executives, between actors and spectators, or among the actors themselves? Assuming that class fractions and class interests could be identified, is the problematic position of actors within the film industry best characterized as a conflict of class?

What must be emphasized is that not all relations of power or conditions of conflict can be or should be articulated in terms of class. In her rearticulation of the Marxist political subject, for example, Chantal Mouffe argues that all social relations can potentially become the locus of antagonism insofar as they construct relations of dominance and subordination.[42] One locus of antagonism that directly impinges upon the construction of actors' subject identities is labor–management relations.[43] Although subjectivity cannot be construed solely on the basis of one's position within the relations of production (a situation that has traditionally given rise to an economically determined class subject), *actors'* subjectivity must be theorized primarily in relation to the production process of the cinema. This is because actors are not simply subjects "in the world," but are defined, at least initially, in relation to a specific form of employment or labor. Their position as laborers establishes an arena of struggle within which the battle of defining subjectivity occurs. As such, issues of labor cut across class lines to position subjects discursively and materially within social relations of power.

This emphasis on labor grows out of and is meant to be consistent with some of the early writings of cultural studies. In *Marxism and Literature*, for example, Raymond Williams remarks that the concept of "labor" became too narrowly conceived in orthodox Marxism. Whereas labor was posited as the primary activity of culture, or as the means by which we (re)produce ourselves and our societies, it more often became equated solely with wage labor, that is, a *specific* form of material production.[44] As a result, Marxists have overlooked the presence of labor in such activities as language and art (often assigning them to the realm of superstructure where they are treated as mere reflections of an economic base), and "the actual work" involved in producing language, literature, and so on is suppressed. The problem, says Williams, is that Marxist theories are "not materialist *enough*" (emphasis added).[45] They should be investigating language (discourse) as labor and all other human activities as part of the "work" of cultural (re)production. Stuart Hall argues further that a materialist theory must encompass some concrete way of thinking the relationship between material and social production "if it is not to desert the ground of its originating premise[:] the foundation of human culture in labour and material production."[46]

Since subsequent work in cultural studies shifted much of its emphasis toward the problematic of cultural reception without an accompanying theory of labor, one of the motivations of this project has been to resituate labor at the heart of cultural theory. Starting with the assumption that the terms of class conflict and the oedipal trajectory at once assume too much *and too little* about how social relations are materially and discursively (re)produced, I shift my investigations of subjectivity toward a notion of *labor power differences* (as opposed to class differences or sexual differences). This concept foregrounds the role of the laboring subject and the struggle over constructing that subject within social, material, and discursive relations of power and representation.

My choice of the term "labor power differences" is meant to incorporate the concept of "dominance" in response to Hall's warning that "to lose the ruling-class/ruling-ideas proposition altogether is . . . to run the risk of losing altogether the notion of 'dominance.' "[47] But the term also signifies an indebtedness to Marx. One of Marx's major interventions into the classical political economy debates over labor was his idea that a worker sells not labor to the capitalist, but *labor power*.[48] As the *capacity* to perform socially useful labor, labor power is placed at the disposal of capitalism in exchange for wages. The value of laboring power, determined by that which is necessary to produce, develop, maintain, and perpetuate the laboring power, thus belongs to the capitalist. Within this structure of social relations, labor power acquires use value (living labor) and becomes a commodity that is transformed into an object of exchange. Labor power remains a "peculiar" commodity in being a value-creating force (i.e., of surplus value) for the capitalist. But, the "diminution of labour-power to the status of an article of exchange effectively means that the worker has simultaneously traded away the positive potential for self-realisation that is inherent within his or her labour-power."[49]

My departure from Marx is marked primarily by the presence of the social subject. Marx's notion of labor power remains individualistic ("The labouring power of a man exists only in his living individuality").[50] Accordingly, the individual's labor power is a condition of being that remains unchanged even though it may be utilized or exploited differentially by the demands of capitalism (as concrete or abstract labor). Through its transformation from use value into exchange value, labor power is also stripped of social agency such that workers under capitalism lose control over the potential embodied in labor. I want to argue alternatively that individual laborers become produced as social subjects through the process of labor, that the terms of subject formation can change even when the basic material conditions do not, and that the *subject's* labor power is neither an individualistic capacity nor a determined effect of capitalist relations.

In a return to the roots of Marx's theory of labor power, I want to foreground the notion of its socially useful *capacity*. Here, before the use value of labor is determined as a commodity, lies the strength and potential of labor power as a political weapon. In other words, it is precisely the use value of (living) labor that becomes the contested ground of struggle within capitalist relations, and it is through the process of this struggle that the subjective dimensions of labor are produced. Within the context of this study, then, labor power is treated as a *relational* concept that addresses the process of subjective struggle within social formations of work. Assuming that capitalists/employers are also "laborers" (to the extent that they perform certain laboring activities), labor power is conditioned by the working subject's position within a hierarchy of labor and by *discourses of labor* that position subjects politically and ideologically as well as economically. A theory of labor power differences is thus concerned with the subject's capacity to control the extent of his or her labor power, the social relations of labor exchange, as well as the concrete conditions of labor that provide the context(s) for such exchanges.

Situating issues of subjectivity within actors' labor history (via a theory of labor power differences) is a tricky business. More often than not, I end up describing *modes* of subjectivity, or various subjective positionalities, rather than the subjectivity of any given actor (as social subject). This is due primarily to the lack of information available on specific actors/stars in relation to specific events. Thus, while I am concerned, like other cultural theorists, with "privileging the subordinate" and giving voice (in this case) to actors' concerns within the labor power differential, few actual voices are heard. Insofar as these individual voices are documented and available, I have tried to include them. Usually the "voices" that are included are those of stars or actors who have assumed some position of authority—the "right" to speak or the "power" to be heard. Even so, these voices are included not so much as a means to authenticate certain individuals (or individual subjects) as to signify various subject positions available for actors as laborers. Moreover, these voices are always already "positioned," as they have already been discursively represented in historical accounts and documents.

One of the major tasks of this project, then, is to map the terrain of actors' labor and subjectivity, to locate the various sites in which actors' labor power and subjectivity is constructed, fought over, and played out. The result is not a conclusive history of actors' labor or even a fully documented account of their struggles. It is rather an *articulation* of the intersecting practices and effects that make up the geography of actors' labor. As Lawrence Grossberg points out, "The notion of articulation prevents us from postulating either too simple a beginning or too neat an end to our story."[51] Articulation involves instead "a continuous struggle to reposition practices within a shifting field of forces . . . by redefining the field of relations—the context—within which a practice is located."[52] Accordingly, in the attempt to manage these shifting forces, historical and theoretical analysis is a process of constructing a map, defining the contours and regions of various practices, and following lines of articulation into spaces where those practices get negotiated, organized, or deferred.

If there are any "lacks" in my own project, they undoubtedly arise out of the difficulties involved in integrating history and theory—a problematic that has long plagued cultural studies.[53] As Richard Johnson notes, "The rooted empiricism of historical practice is a real liability often blocking a properly cultural reading."[54] This is why the historical data I present sometimes take on a life of their own, overpowering the theoretical framework and tending toward a linearity and truth value that begs for more deconstruction. At other times, however, the theory that drives the engine of my historical analysis is alternately vague, long-winded, or not scrutinizing enough. I can only hope that the problems of this text—and there are many—are viewed not solely as "lacks" but also as "conditions of possibility" for constructing more detailed and ever better maps for theorizing and historicizing subjectivity from a labor perspective.

2 / The Subject of Acting

Labor and the Commodity Form

The concept of *shifting, fragmenting, and binding,* which stems from Marx's theory of commodity fetishism and which has been used in cultural studies to describe various forms of institutional oppression,[1] is particularly useful in understanding the ways that studios positioned actors as social subjects during Hollywood's Golden Era. By virtue of their labor power advantage, studio executives were able to determine the ideological boundaries of labor discourse as well as the economic boundaries of labor conditions and commodity exchange, thus limiting actors' ability to control the terms of their subjective representation. In transforming labor into commodity form, film studios first of all shifted emphasis and visibility from production to exchange. This shift masked not only the process of labor but also the antagonistic relations that existed within the sphere of production. In addition, consumer attention was shifted toward a commodity form that concealed the source of its value by "behav[ing] as though value were a property of the commodity" itself.[2] Indeed, in the case of actors, the fruits of labor produced a commodity that was particularly rich in surplus value. Even though the image or star icon was dislocated from the sphere of production, its representational form appeared to capture "the real thing," thus providing a strong source of fetishistic attachment with which to link the consumer to the actor's body in the sphere of circulation. Since this fetishization translated into money at the box office, the studios were more concerned with promoting star images than with acknowledging or improving the working conditions of actors.

Fragmentation entered this process in a variety of ways. On a structural level, the star system established a hierarchical division of actors' labor that allowed the studios to maintain economic and political control over the acting profession. The various divisions, based on cost and use value, were also necessary for efficient production. Stars represented the smallest group because they were so costly, but due to their use in regulating the industry's product and drawing audiences into the theaters, they were also extremely profitable to the studios. Character actors and bit players made up a somewhat larger group since they could be used repeatedly in films at lower wages than stars. Screen extras represented the largest group, sometimes estimated as high as 90 percent of the acting profession.[3] Extras could be paid minimal wages, and their chances of repeated use by a studio often depended upon their willingness to work below current wage standards and outside established labor guidelines.[4]

The practice of typecasting fragmented the acting profession further. Within each level or division of labor, in other words, actors were categorized according to social types based on race, age, sexual stereotype, and so on. Such typecasting not only fragmented actors' labor power (by limiting their range of performance and preventing the full potential of their skills), it fragmented actors' bodies as well. Often this fragmentation was sexualized. As Hortense Powdermaker observed in her anthropological study of the film industry,

> An actress becomes known for her comely legs, and these are accented in every picture. Another one is known for her bust; still another for her husky, sensuous voice. So obvious is the use of actors as sexual symbols that in a major studio a handsome star is colloquially referred to as "the penis."[5]

Not surprisingly, actors performed, or were coerced into performing, sexual labor. Young starlets, for example, were promised acting parts or screen tests in return for their sexual labor on the "casting couch." The labor that actors performed on their own bodies was also sexualized. Since actors were expected to maintain a certain look or body weight in order to achieve and sustain the success that accompanied stardom, anorexia and breast implants for women and rigorous body building and hair implants for men were not uncommon.

Thus, the process of fragmentation worked not only to differentiate star types in the sphere of exchange, but to create and sustain differentiated labor in the sphere of production. The overall effect of this fragmenting process, however, was to emphasize *differences* among actors. By creating and reinforcing individual differences of salary, opportunity, skill, personality, sexuality, and social type, the process of fragmentation established an "isolation effect" that placed barriers between actors and forced them into competition with each other for studio attention.[6]

The fragmented state of the acting profession gave studio heads the power to bind actors into a passive community of workers. A constant pool of unemployed and underemployed workers (mostly extras) made it possible for studios to reduce labor dissension.[7] The promise of moving up in the star system hierarchy kept hopefuls in line, while the fear of plummeting to the bottom was used to keep employed actors from challenging their employers and complaining about exploitative labor practices. Though actors were indispensable to the production of the commodity form, cooperation with studio policy was thus a precondition of achieving a livable salary and job security.

Binding was reinforced through an ideological discourse of the family that denied labor power differences, yet functioned according to a model of "paternal tyranny" whereby "love, honour and obedience were venerated and rewarded, while the neglect of them earned self-righteous retribution."[8] Actors thus found it difficult to organize their collective power. Even when they managed to form political alliances across internal divisions of labor (or with other labor groups) they had to fight against a double stronghold of management and family that threatened them with dismissal and branded them as ungrateful children. The combined effect of the shifting, fragmenting, and binding process was to limit actors' power as active social agents, constructing them as passive workers and producing them as passive images.

This is not to suggest that actors' subject positioning can be explained as a simple reflex of the economy. As Jane Gaines states,

> Although the functioning principle of capitalism is the complete organization of social and economic life around the commodity, human needs vacillate and don't always line up with commodity use-values. . . . In political terms, this means that the social

and economic arrangements supporting capitalist production are in constant danger of coming undone.[9]

This situation is intensified by the fact that actors are caught between the forces of production and consumption, between bodily labor and commodified image. A gap exists between an actor's use value (labor) and exchange value (image). Even if a stabilized image could be established in the sphere of circulation (which was rarely possible), it would not guarantee a "stabilized" or obedient worker in the sphere of production. Likewise, although the economic and psychological effects of the star system's fragmentation may have caused actors to feel alienated from their fellow workers and from their own labor power and bodies, it could not entirely suppress or control actors' resistances.

While actors negotiated the difficult terrain between representation and self-representation, the studios attempted to appropriate the actors' labor into an economic and ideological system of profitability, mobilizing their "free consent" to establish hegemony over actors' labor and subjectivity. As Tony Bennett explains,

> It is not enough that the worker should be reproduced as someone capable of work and socially dependent on capital; he or she must also be produced as a subject of an ideological consciousness which legitimates the dominance of capital and the subordinate place which he or she occupies within its processes. . . . They must be induced to "live" their exploitation and oppression in such a way that they do not experience or represent to themselves their position as one in which they are exploited.[10]

But the lines between free consent, self-determination, and exploitation become difficult to sort out. Some actors, for example, were in a position to reap considerable rewards from the star system. The exorbitant salaries and international fame that accompanied stardom in some cases outweighed, and even diminished, exploitative labor conditions. Although wary of the consequences, actors lower in the hierarchy also stood to gain. This may explain why Alexander Walker found R. D. Laing's studies of schizophrenia apropos in describing the star as "a person [who] 'cannot make a move, or makes no move, without being beset by contradictory and paradoxical pressures and demands, pushes and pulls, both internally from himself, and externally from those around him.' "[11]

Barry King suggests, in addition, that some actors walked a tightrope in balancing the studios' demands with their own desire for control. Since performance confronted actors as a fact of employment, some attempted to utilize it in a way that met both their own needs and the needs of the institution. For example, "actors seeking to obtain stardom [would] begin to conduct themselves in public as though there [was] an unmediated existential connection between their person and their image."[12] While this strategy complied with the studio expectation that actors would develop "personalities" for purposes of public visibility, actors who initiated this process could claim some degree of control over the form their representation took. Similarly, actors attempted to control the details of their film performances in order to claim credit for their work. When successful, this strategy created the identity of a "good dramatic actor" as opposed to a "popular Warner Brothers star," thus allowing the actor (instead of the studio) to receive recognition for his or her own labor.

At the core of actor–management struggle, then, is the issue of how and how much labor is to be attached to the image. The actor, says King, is placed in a paradoxical position in relation to this issue: "While film increases the centrality of the actor in the process of signification, the formative capacity of the medium can equally confine the actor more and more to being a bearer of effects that he or she does not or cannot originate."[13] Given the displacement that occurs in cinematic practice from production to signification, the labor of performance becomes less the provenance of actors and more the property or product of the studio. This paradox, concludes King, causes the "persona" to become a site of struggle within the hegemonic discourses of the cinema: "A potential politics of the persona emerges insofar as the bargaining power of the actor, or more emphatically, the star, is materially affected by the *degree* of his or her reliance on the apparatus (the image), as opposed to self-located resources (the person) in the construction of the persona."[14]

This explanation is adequate to the extent we recognize that the persona is not a predetermined, easily defined entity, but a construction that results precisely from the struggle over the image–labor relation. But, given this, it might be better not to focus solely on the persona, but to see "the persona," as well as "the image" and "the person," as terms that arise out of the discursive struggle between labor and management. The studios, for example, worked toward separating notions of image and person as a means to construct and establish exclusive control over a coherent, salable persona. That is, by detaching the image from the person, the studio could reconstruct the relation between the two into a unified subject position called the "persona." Changing the names of actors was one way of establishing this control.[15] By erasing an actor's previous "identity" (name, personal history), the studio could create a new image and identity. An agreement of sorts was made: in return for the actor's physical body (as bearer of an image), the studio would attempt to generate for her or him the wealth and social prestige of star ranking.

The studios' power of creation and naming assured them economic and ideological, if not legal, ownership over the actor's body. Indeed, as King points out, "The established policy of building stars from inexperienced players under the studio system, can be seen to contain an element of fabricating subordination among potential stars."[16] By freely consenting to this process of fragmentation and individuation, actors lost political autonomy and forfeited some control over defining their own subject identities. It is true that one's desire for the "coherent subjectivity" of the star persona (or what King has called the single "transfilmic star personality image") was "in line with the star's economic interests, since the further he or she [could] enforce such an equation [between the person and the image] the greater his or her irreplaceability and bargaining power."[17] But regardless of the prestige and control that stardom might afford the individual actor, the structure of labor relations itself was not challenged, and the labor power differences between actors and management remained intact.

An actor's signature meanwhile permitted his or her "image" to become the legal property of the studio. The Screen Actors Guild reports that between 1937 and 1946, the number of players under long-term contracts to Hollywood studios varied between six hundred and eight hundred.[18] The usual contract ran for seven years, with the studio having the right to take up the option of an actor's services after six months or a year with an increase in pay. If the studio did not take up the option or wished to fire the actor (with or without stated causes), the contract was terminated. The actor, on the other hand, could not legally break the contract under any circumstances. By individualizing actors' labor power under separate contracts, producers could regulate the division of labor under the star system and enforce the use of typecasting. Since most contracts stipulated that actors must accept the roles "offered" to them, most actors found themselves playing the same type of role over and over again. The repetition of roles reinforced the illusion of a unified persona. To break from this illusion and to challenge studio regulation of the persona often resulted in suspension from the studio without pay, and, upon readmittance, the forced acceptance of even less desirable roles.[19]

The typical contract gave a studio exclusive right to "photograph and/or otherwise produce, reproduce, transmit, exhibit, distribute, and exploit in connection with [a] photoplay any and all of the artist's acts, poses, plays and appearances of any and all kinds."[20] In addition, the studio had exclusive right to determine who else might be allowed to use the actor's image for advertising or commercial purposes. With very few exceptions, actors had no right to their images and no control over how their images were exploited, divided, or transferred. It was also not unusual for actors to be unaware of how their likenesses were being used. "An actress," says Gaines, "might be shocked to see her image reproduced in conjunction with products as diverse as Auto-Lite car batteries or Serta mattresses and box springs, but there was little she could do about it"—as Kay Francis discovered when she found herself advertising Compo shoe soles in a 1933 issue of *Photoplay* magazine.[21]

According to King, however, the image remained "a slippery commodity" due to the conflict that existed between the legal and physical aspects of image ownership. Although the exchange of labor granted studios the right to use or police an actor's image in ways that were profitable to them, the image was not "property" in the usual sense and could not be entirely owned:

> Whilst the contract for the employment of the star closely stipulates the limits on the utilization of the image . . . nevertheless it remains factually the case that the star is ultimately the possessor of the image, because it is indexically linked to his or her person.[22]

In effect, adds King, a legal monopoly was confronted with a physical or "natural monopoly" in a bargaining relationship. Although this gap between "the image" and "the person" cannot in itself account for the antagonisms between labor and management (which appears to be the implication of King's argument), it is true that the studios' power did not go uncontested; they were forced to bargain with actors over rights to the image.

Thus, although studios exerted considerable control over the terms of actors' contractual obligations, contracts can also be seen as a means for the two parties to work out their conflicts over the image. As Gaines notes,

> The studio used the contract to secure on-screen continuity through provisions for illness and vacations, wardrobe fittings, tardiness on the set, absences, photographic sittings and re-takes. . . . Stars retaliated with riders modifying wardrobe requirements, stipulating screen billing order and typeface size on the credits, and mandating the number of closeups per picture.[23]

Yet the contract remained a paradoxical document. Although it created an imaginary relation of fair exchange between legal subjects, the contract spelled out the terms by which one party would be able to transform the labor and representation of the second party.[24] In addition, the contract exposed the constructedness of the star image at the same time it created the possibility for coherence. This is surely one reason that the studios responded so harshly when actors challenged their contracts. Such challenges ran the risk of disrupting the coherent images that studios wanted to present to the public.

Fortunately, from the studios' point of view, they did not have to rely solely on contracts to create a sense of coherency. Whereas contracts functioned within the sphere of production, realist films (and their accompanying publicity) reinforced the illusion of coherency in the sphere of exchange. As John Ellis has noted, star images function(ed) as "an invitation to the cinema."[25] Because star images were composed of character fragments, chunks of "real life," still photographs, and disembodied radio voices, the cinema's combination of voice, body, and motion promised to reveal the completeness of the star image and "the mystery of the star's essential being."[26]

Discourses of realism worked toward unifying not only actor and image, but also actor and character. While suture functioned in the case of spectators to produce identification with ego ideals, the production of actors as those ego ideals sutured actors into a discourse that naturalized their labor as performance and linked their performances to narrative identity. According to these conventions of naturalism, stars did not work. Though some stars were distinguished for their acting ability, it was widely thought that most stars owed their success to their personalities or photogenic qualities. In 1933, *Variety* reported:

> In hopes of finding featured talent that can be eventually developed into star material, all majors are on the widest search in picture history for new *faces*. Scouts are going into all fields which might possibly yield screen *personalities*.[27]

Viewed as "personalities" as opposed to "laborers" in a realist economy of representation, they merely displayed the qualities they possessed or the personalities that were packaged for them by others.

Character actors and supporting players fared much better in this regard. Generally considered the most talented of Hollywood actors, they were thought to carry the real burden of acting and were often relied upon to conceal a star's lack of ability. Supporting actors, in other words, were valued more for the labor they performed; they were viewed as "impersonators" (actors possessing impersonatory skills) rather than as "personalities."[28] Not surprisingly, studios spent less time and energy selling the images of character actors to the public. Hollywood dealt in personalities; the less an actor was "labor-identified," the more that actor was promoted as the Hollywood ideal. In spite of this, or perhaps because of this, some actors chose the type-casting of character acting and enjoyed a professional reputation, career longevity, and financial durability not experienced by the more transitory stars.[29] In contrast, the work of Hollywood extras never established the continuity or dramatic depth of the work of other actors. Though they were sometimes used to perform unusual tasks (e.g., daredevil stunts or accordion playing),[30] extras were used in films predominantly as objects to dot a landscape (hence the term "atmosphere players"), and their filmic presence required little or no skill. Unlike stars, however, their objectified state carried no power. Extras were neither personalities nor impersonators, and were defined more by their unemployment than by their employment.

In addition to their objectified status in films and their subordinate status in the labor–management hierarchy, actors were also positioned as "other" in the Hollywood labor community. A common cliché in Hollywood, reported Powdermaker, is that "there are three kinds of people— men, women, and actors."[31] Actors held the dubious distinction of being the only production workers who were visibly present in the industry's products. Technical workers who considered themselves to be the true creative artists of the industry, yet were paid (on the average) lower wages and received less recognition, often resented being upstaged by actors. Stars, in particular, were perceived as a privileged class of workers that was undeservedly pampered by the front office. As a consequence, their complaints about labor practices were not taken very seriously. Stars were branded instead as "immature, irresponsible, completely self-centered, egotistical, exhibitionist, nitwits" who held up production or created unnecessary problems on the set.[32] Even though the various occupational groups would have benefited from a united workers' front, actors by and large received little support, either collectively or individually, from other labor groups. Actors were resented and alienated by other workers in ways that supported the dominant discourse of labor.

Studios thus relied on a variety of labor power differences, internal and external to the actors' ranks, to maintain their own labor dominance. Through the process of shifting, fragmenting, and binding, the studios were further able to create a discourse of stardom that invited the actors' free consent and that was effective in reducing challenges to this dominance. Individual actors who did insist on contributing to their own image construction were regarded as a nuisance, but could be dealt with through contract negotiation, suspension, or dismissal. When actors struggled for collective self-representation and challenged exploitative labor practices, however, more severe measures were instituted to deny them a voice as political subjects (see chapter 3). As actors struggled for the right to define and control their own subject identities as laborers, the studios struggled to position actors as passive objects of display, fragmenting their labor power into institutionalized categories of image, performance, and profitability. Given the conflicts over systems of representation and issues of self-representation, the actors' struggle for definition was difficult.

The Union Question

An understanding of actors' subjectivity requires not only an investigation into labor–management relations, but an investigation into actors' shifting perceptions of themselves in relation to their work and to the cinematic institution in general. It requires, in other words, an understanding of the unifying principles around which diverse groups of actors united or formed "unions" (in the broadest sense of the term), and the way in which fragmented aspects of subject identity cohered in relation to these unions. Since the fragmented condition of subjects always tends toward unity, whether by conscious or unconscious means, whether by free consent or active resistance, the significance of subject unity lies in its political utility. For the unifying constructs of subject identity not only determined how actors perceived themselves or were perceived by others; they directly influenced relations of power in the industry and affected actors' strength as a bargaining unit.

In undertaking such an analysis, I tread upon the terrain of the "collective subject," a theoretical concept that has been all but avoided in film studies (and most Marxist criticism as well) due, I think, to its suggestion of a mass consciousness or collective will. Viewed from the perspective of labor, however, the collective subject—or a collective subject identity—takes on more useful and politically specific connotations. As Raymond Williams notes in *Marxism and Literature*, collective subjectivity involves a process of "conscious cooperation" or collaboration. It is a "case of cultural creation by two or more individuals who are in active relations with each other, and whose work cannot be reduced to the mere sum of their separate individual contributions."[33] In an attempt to distance Marxist cultural theory from a bourgeois notion of the individual, Williams stresses the "trans-individual" nature of the collective subject whereby we can discover "the truly social in the individual, and the truly individual in the social."[34]

In his essay "What is Cultural Studies Anyway?" Richard Johnson argues more forcefully for the need to address the collective dimension of subjectivity. Within poststructuralist theory, he says, "there is no account of . . . *the subjective aspects of struggle*, no account of how there is a moment in subjective flux when social subjects (individual or collective) produce accounts of who they are, as conscious political agents, that is constitute themselves, politically."[35] Thus, what cultural studies must take up is an investigation of how social movements or groups "strive to produce some coherence and continuity." It must engage, he argues, in a "post–post-structuralist" account of subjectivity that returns to and reformulates questions of struggle, "unity," and the production of a (collective) political will. This involves, most importantly, a theoretical notion of the "discursive self-production of subjects, especially in the form of histories and memories"[36]—and, I would add, everyday practices.

Thus, from the theoretical perspective of labor power differences, "collective subjectivity" refers to the process that laboring subjects undergo in forming, maintaining, or protecting a collective sense of identity. Since the social relations involved in this process follow no internal logic nor create inevitable results, the project at hand must "abstract, describe and reconstitute in concrete studies the social forms through which [actors] 'live,' become conscious, sustain themselves subjectively."[37] It must locate the specific configurations of actors' collective subjectivities, which arise out of a history of struggle and labor power differences, and analyze the ways in which knowledge and experience of these collective notions are discursively produced and materially lived within the shifting context of social relations in Hollywood.

The contours and spaces of actors' subjectivity are often difficult to determine. In the earliest years of cinema, for example, a coherent or unified notion of screen acting did not appear to exist. Actors from vaudeville and the stage became part-time "picture performers" to pick up a few extra dollars during daytime hours. Or, within a motion picture company, employees who performed other duties might be asked also to "pose" for the camera.[38] By the 1910s, as the industry sought to legitimate the new entertainment form among the middle and upper classes, screen acting increasingly became defined as a specialized skill, and discourse about the acting profession began to differentiate between the theater and motion pictures. As Richard deCordova has noted, there emerged "a sort of struggle between photographic and a theatrical conception of the body, between posing and acting."[39] Distinctions were made between the live, vocal performances of stage acting and the type of acting required to create the phantom images of the silent cinema. Although film producers often played up an actor's stage experience as a way to legitimate his or her professional existence (and the film industry's existence in general),[40] actors became part of the ever-widening discursive gap between stage and screen.

Material differences also affected actors' notions of themselves and their profession. Screen acting, for example, differed not only in terms of craft, but in terms of the institutional context. From the film industry's beginning, screen actors encountered different working environments and labor power relations in the studios than stage actors encountered in the theater. In the latter, where employee–employer relations were stabilized and ownership was concentrated in the hands of a few, actors suffered a number of abuses. Alfred Harding, a historian of early stage labor, states that, in contrast, motion pictures offered lucrative and relatively stable employment conditions without the accompanying abuse by management: "There was still so much money to be made from the making, booking and exhibiting of motion pictures that the money to be gained by rigging the actors, considerable as that sum would have been, was a mere drop in a capacious bucket."[41]

As the film studios moved their operations to California, stage and screen were separated even further, and the "difference" of screen acting intensified. Once the Hollywood star system became more firmly established and divisions within the talent group were intensified, a greater distinction between high-ranking and low-ranking actors also emerged. Before World War I, screen actors formed their own labor associations, but these groups were primarily social or benevolent organizations and "had little interest in or orientation toward industrial relations."[42] Later, high-ranking actors formed the Screen Actors of America, and atmosphere and bit players belonged to the Motion Picture Players Union (MPPU). Although neither group was a radical political body, both had obtained a charter from the American Federation of Labor and established an orientation toward industrial relations.

These labor groups were challenged in 1919 when Actors' Equity Association, the political body formed by stage actors in 1913, sought jurisdiction over Hollywood.[43] Although Equity had hoped to "penetrate Hollywood peacefully," it was met with resistance, and it took several months of negotiation before the existing screen actors' unions agreed to acknowledge Equity's jurisdictional rights. By this time most screen actors were bypassing the stage and beginning their careers directly in the cinema. As Murray Ross explains, they did not know the history of labor struggles in the theater and were not interested in the stage actors' problems.[44] Equity also threatened the screen actors' professional autonomy. Because Equity had moved into Hollywood so shortly after winning a major battle with Broadway theater managers, many actors believed that the union's interest in controlling Hollywood merely stemmed from its desire to strengthen its home bargaining position.[45]

Equity members, however, argued that their actions were motivated by a spirit of collectivity and should not be interpreted as opportunistic or divisive; their goal was to protect Hollywood actors from the sort of exploitative conditions that had occurred in the theater. Although "a general survey of conditions affecting motion picture actors on the Pacific Coast revealed that at that time there were prevalent remarkably few of the abuses which had driven the dramatic actors to organize," members of Equity wanted to offset any advantage that management might gain.[46] The need for a "strong and watchful organization" was not apparent to Hollywood actors because the motion picture industry was still young and had yet to develop entrenched relations of labor–management power. But, as Equity noted, "The industry was beginning to crystallize." If protective measures were not taken soon, actors would witness "the consolidation of the field in the hands of a few strong men."[47]

During its first few years in Hollywood, Actors' Equity continued to monitor the situation without taking action. Equity felt that "the majority of motion picture actors were not yet ready to be organized" even though abuses against actors were beginning to mount.[48] In addition, unionization of the sort Equity had achieved in the theater was not yet possible, because no official employer bargaining unit existed. It was not until the Motion Picture Producers and Distributors Association (MPPDA) was formed in 1922 that Equity sought to negotiate its first standard contract.[49] Will Hays, head of the MPPDA, "noncommittally agreed to consider the request," but the matter was apparently ignored. When the Association of Motion Picture Producers (AMPP), the labor branch of the MPPDA, was formed in 1924, Equity approached Hays again. But Equity's request for a standard contract, closed shop conditions, and studio recognition of the actors' union was flatly rejected.[50]

Equity members did not push the issue further because they were unable to garner enough support in the screen community. In addition, some of the labor practices that Equity had protested against were temporarily discontinued when Joseph Schenck, president of the AMPP, intervened on the studios' behalf.[51] Thus, although interest in Equity "had been high," many screen actors thought the newly formed AMPP had responded adequately to their needs. Equity backed off, allowing their recruitment drive to come to a "virtual standstill," but it refused to recede into the background.[52] In a statement issued to the press, the union declared:

> Equity wants it understood that it is not abandoning its Los Angeles office and that it is not contemplating any such action. . . . it is in Los Angeles and the motion picture field to stay, and will be there strong and vigorous long after these short-sighted actors and actresses have become dusty shadows on rolls of celluloid in somebody's storage warehouse.[53]

Equity was clearly becoming impatient with the naive and uncooperative behavior of their fellow actors in Hollywood even though the union was committed to protecting all members of the same profession.

The less than harmonious relationship between stage and screen actors, however, was not simply a matter of naïveté. Equity's attempt to protect and educate Hollywood actors also involved a control over and redefinition of screen actors' subject identity. The identity of "actor" (versus "screen actor") threatened their professional autonomy by denying the specificity of their labor and their relation to Hollywood. Screen actors were, moreover, inclined to view themselves as "picture personalities" or members of a "studio family" rather than as "industrial workers." As producer Milton Sperling observed,

> In those days in Hollywood, studio loyalty was a factor of your life. If you were a Warner employee, or a Fox employee, or a Metro employee, that was your home, your country. . . . You played baseball against the other studios. You had T-shirts with your studio's name on them. It was just like being a *subject*, and a patriotic subject at that. People who lived and worked beyond the studio walls just didn't belong, and you were prepared to fight them off, like the Philistines.[54]

But the screen actors' failure to recognize or confront the broader implications of Equity's efforts carried a high price. In their desire for an autonomy and subject identity based on film specificity, screen actors repeatedly sided with their motion picture employers and rejected the labor history and bargaining experience of their fellow workers in the theater.

The position chosen by screen actors (and fostered by the studios) left them more vulnerable to studio domination. By 1926 the Hollywood labor situation had undergone some fundamental changes. Though the major studios still maintained an open shop policy, they had signed the Studio Basic Agreement with the craft unions and were gradually becoming involved in the process of collective bargaining and collective negotiations. Studio heads realized that similar measures would be necessary if they wished to maintain their control over the creative talent groups. Thus the studios began to make certain concessions in the hopes of appeasing the demands of talent groups while forestalling their unionization. These concessions (e.g., a more equitable distribution of work for extras through the establishment of the Central Casting Corporation) were designed not only to undercut Equity's influence in Hollywood, but to discourage screen actors' identification with (unionized) actors from the stage.

As part of this new managerial approach, the producers created the Academy of Motion Picture Arts and Sciences. The Academy was made up of five branches representing the major divisions of motion picture production: producers, directors, writers, actors, and technicians. According to the original charter, the branches were to be equally represented on the Academy's board of directors, and each branch would elect an executive committee by democratic process to function as its governing voice in labor negotiations. In an effort to make the Academy a prestige organization above the status of a labor union, membership was by invitation only and based on one's distinguished accomplishments in film production. According to labor historians Louis B. Perry and Richard S. Perry, the Academy's structure was particularly attractive to major screen stars who were growing uncomfortable with Equity's attempts "to organize and control from 3,000 miles away."[55] Unlike the theater union, the Academy recognized stars as members of a white-collar profession who should be treated on the basis of individual artistry. The Academy's "method of selection, however, kept the control of the organization in the hands of a few, so that it took on many aspects of a company union."[56]

Actors' Equity was suspicious of the Academy's commitment to labor issues and believed that its formation was "calculated to give Equity a final blow."[57] Thus, when the AMPP announced a 10-25 percent reduction in salary for nonunion labor in 1927, Equity stood ready to challenge the Academy's stated commitment. The screen actors' response to the situation repeated a familiar pattern. Feeling betrayed by the producers' association, they turned to Equity for assistance. But when the newly formed Academy protested successfully against the salary reduction (by convincing producers to consider the merits of each individual case), actors placed their allegiance with it. Thus Equity was once again forced into inactivity. Its presence continued to serve as a "deterrent to unlimited aggression on the part of producers,"[58] but producers still assumed the right to speak for actors through the benevolent auspices of the Academy.

The question of "a voice," of making oneself heard, took on an added significance in the battle between labor and management when the arrival of sound accentuated the voice as a material site of struggle. As Walker notes, the "economic dislocation" caused by the switch-over to sound technology was also accompanied by a "human dislocation."[59] Articles in the trade press capitalized on "scare stories" and predicted an apocalyptic outcome for even the most well established silent actors. Nervousness about learning new techniques caused some motion picture stars to enroll in voice production schools or to go to Broadway to establish themselves as stage actors. But, according to Walker, "The more insecure the talkies made these highly-priced and troublesome people feel, the better a front office liked it."[60] Producers learned quickly that while the sound crisis was stirring antagonisms between stage and screen actors, it was also increasing their control over the labor force. Producers used the crisis as an opportunity to cut the escalating salaries of stars. Since it was economically advantageous to purchase talent that was already developed, they also brought in "proven voices" from the stage at a cheaper rate. The competition and feeling of insecurity that this created among Hollywood stars subsequently persuaded them "to take cuts, or resign at a lower figure, in order to hang on to their stardom."[61]

But although the employment of stage actors provided a quick fix for the studios, producers were opening the door for the subversive potential of the voice. Of the approximately twelve hundred stage players who migrated to Hollywood to appear in the talkies, nearly all were members of Actors' Equity. These actors were accustomed to an Equity shop policy in New York theaters and had experienced firsthand the sorts of improvements that Equity had been able to obtain from theater managers. By 1929, 70 percent of all actors in talking pictures (including screen actors who joined locally) were Equity members. The membership was active, filing complaints about studio working conditions at Equity headquarters and calling for all-Equity casts.[62] With such unprecedented support from screen actors, union officials thus decided to make another stand for an Equity shop policy in Hollywood.

During their struggle to obtain the voice of effective self-representation, actors' groups underwent a series of realignments. First and foremost, stage and screen actors developed more harmonious relations. Some film actors, particularly the higher-ranked ones, continued to resent the presence of the theater in Hollywood and were suspicious of the union's attention to the newly arrived stage players. The fact that Equity had suspended several leading actors from the union (for violating Equity regulations) only added to their antagonism.[63] But the majority of actors from the lower ranks (character actors and bit players) welcomed the bargaining position that Equity could help them achieve. This vote of confidence and solidarity was forcefully expressed at a rally of Equity's members when, upon adjourning, the crowd of actors sang the song first used in the stage actors' theatrical strike of 1919, "All for One, and One for All."[64] But while these screen actors forged a unified front, of sorts, the terms of collective subjectivity remained fragmented.

It became increasingly apparent that the major split within the acting profession during this period was no longer based on medium specificity (i.e., stage versus screen), but on a hierarchical notion of actors' labor. Equity's active membership came from the lower ranks; these stage and screen actors defined themselves as workers, and the organization itself was structured along trade union lines. The major motion picture stars resisted the definition of actors as workers. They also feared that the union drive would cost them the status and power they had worked so hard to achieve. A number of them belonged to both Equity and the Academy, but as labor historians Perry and Perry point out, they were "not likely to quit the Academy in favor of a union until they found a lack of good faith in the former and were ready to consider themselves as workers in need of a labor organization rather than members of a professional group who were above organization."[65]

Gaining the loyalty and commitment of prominent actors was essential to Equity's overall success. But the stakes and issues were vastly different for stars than they were for other classes of actors. Whereas the distinguished stars could arrange contracts that guaranteed high wages and specified certain favorable working conditions, most of the industry's actors—especially screen extras—were not in a position to bargain. If they spoke out against abuses they were seldom reemployed at the same studio; and since they were never sure of continuous work, most actors kept quiet. When these actors did obtain work, they were often forced to accept contracts that were "hopelessly vague and inequitable" and essentially amounted to "tak[ing] the casting-director's word."[66] Union support from their prestigious and steadily employed colleagues would thus give them a bargaining edge that their mere numbers could not ensure.

According to a report in the *Nation*, however, producers were "resorting to every conceivable device to break the spirit of the actors."[67] They tried to undercut the union drive by offering actors tempting non-Equity contracts. Those who refused had their names passed on to other studios, where they would find it difficult to obtain work.[68] Producers also relied on the local newspapers to further their antiunion crusade. Lists of non-Equity members, for example, were published in the local press to help studios "make their hiring decisions." Both the *Los Angeles Times*, a notoriously antiunion publication, and the *Los Angeles Examiner*, owned by William Randolph Hearst (a major stockholder with MGM), printed lengthy editorials against Equity. In one, stars were warned that an affiliation with Equity would turn them into blue-collar workers, because Equity had "placed itself in line and agreement with stagehands, ditch-diggers, janitors, iron-molders, and such."[69] Meanwhile, the local press printed interviews with several of Hollywood's top stars (Lionel Barrymore, Louise Dresser, Marie Dressler, John Gilbert, and Norma Talmadge) who praised the producers for attempting to negotiate fair—that is, non-Equity—contracts.[70] These stars preferred that labor negotiations be handled by the Academy, an organization they had helped to create.

Equity once again miscalculated the unity and strength of screen actors. Although more than two thousand Equity members had turned down nonunion contracts, producers had held on to enough of the important actors to maintain continuity in production as well as a bargaining edge. The absence of prominent actors from the bargaining table, and the lack of political support, weakened Equity's position. Organized labor also retreated from the scene. Although several Hollywood craft unions had pledged their moral support and, in some cases, even their financial support to the actors' cause, they now refused to call sympathetic strikes.[71] In addition, fighting among the internal ranks of Equity resulted in the union coming up empty-handed. Although the producers, at one point, had consented to 80 percent Equity and 20 percent nonunion labor in all casts, indecision and delay among Equity officials caused producers to withdraw their offer.[72]

The breakdown in Equity leadership, the conflicting interests between high-ranking and low-ranking actors, and the lack of outside assistance resulted in an overwhelming defeat. The acting profession was now more vulnerable and fragmented than ever. And, left with no other option, actors hurried to accept contracts on producers' terms. As workers, actors once again were forced to define themselves individually—rather than collectively—in relation to producers. The political gap between high-ranking and low-ranking actors intensified, and the more harmonious relationship that had developed between stage and screen began to dissolve (or at least it became less consequential). Now that all actors in Hollywood were forced to deal directly with motion picture producers, the issue of "film specificity" reemerged as the organizing principle of actors' subject identity. But this time the voice of theater was silenced on producers' terms. This would remain the case until actors were able to forge another, more collective, discourse of labor to define themselves differently.

3 / The Politics of (Self-) Representation

According to Marx, the inevitable consequence of relations between labor and capital is crisis. This is due not to some inherent "logic of capital," as some Marxist critics have interpreted Marx's claim, but to the social relations of power within the capitalist mode of production. As Eric Lichten notes in *Class, Power and Austerity*, labor is not merely an *effect* of capital but a barrier to capital accumulation that often takes the form of a conscious resistance by workers who seek greater autonomy within the capitalist mode of production.[1] This is another way of saying that the relation between labor and capital is a historical one. However inevitable this relation might appear, we cannot know the specific circumstances of its crises until we examine the history of labor power differences and the material and discursive struggles that arise out of these social relations.

A number of theorists have noted the advantages of examining social relations of power during periods of crisis. In *Policing the Crisis*, for example, Stuart Hall et al. argue that a crisis exposes the process of free consent used to maintain hegemony:

> If, in moments of hegemony everything works spontaneously so as
> to sustain and enforce a particular form of class domination while

rendering the basis of that social authority invisible through the mechanisms of the production of consent, the moments when the equilibrium of consent is disturbed . . . are moments when the whole basis of political leadership and cultural authority becomes exposed and contested.[2]

Lichten further suggests that crisis is a moment of great transformation, "a moment of alternative possibilities—a moment of many possible histories."[3] In a crisis situation, the relations of power become more apparent and make the contested terrain of struggle more visible. Emergency or coercive tactics taken by management, for example, reveal the extent of managers' domination over workers, whereas the strategies initiated by labor test labor's strength or weakness as a destabilizing force. As the political and economic stakes of labor power differences become exposed, groups struggle to maintain or redefine their internal relations of power as well. The events that precipitate crises create rallying points for this purpose, forcing groups to (re)position themselves around certain economic and political issues. Through this process, crisis events can threaten group identification and solidarity or provide opportunities for the formation of new identities.

This struggle over identity and positioning ultimately hinges on the question of a subject's individual and collective relation to labor power differences. The way in which identification is constructed determines the boundaries of material and discursive struggle over labor power and subjectivity. Thus, as I argue here, the history of actors' labor struggle is also, necessarily, the history of their subject formation. The conflictual aspects of subject formation enter into labor struggle, influencing and directing political actions, while these actions, in turn, mold and define the terms of subject identity. During moments of labor–management crisis, the various political and psychical components involved in the process of subject formation become foregrounded in a rather dramatic way and offer insights into the choices that are made and the changes that occur within this arena of struggle.

"Crisis" is a term that has long been associated with Hollywood. As Thomas Schatz notes, "The Hollywood studio system, with its factory-based mode of production and division of labor and its distinctive relations of power, not only permitted but virtually demanded a degree of struggle and negotiation in the filmmaking process."[4] In her study of Hollywood's studio-era culture, Hortense Powdermaker argued that studio executives actually turned this situation to their advantage by maintaining an atmosphere of continuous anxiety. Through a corporate practice of "crisis management," political inertia could be induced while crisis itself was averted. Thus, she observed, there is "a kind of hysteria [in Hollywood] which prevents people from thinking, and is not too different from the way dictators use wars and continuous threats of war as an emotional basis for maintaining their power."[5] Consequently, the atmosphere of anxiety "prevent[ed] an awareness of the factors which call[ed] it forth" and kept workers from doing something about it.[6]

While my analysis of actors' unionization in chapter 2 suggests that actors in fact made many attempts to shift the balance of labor power between actors and studio executives, it also leaves the misleading impression that actors were never successful in gaining control over their own subject identities. Perhaps as long as producers called the shots and were able to control the atmosphere of crisis (e.g., the uncertainty over contracts, the lack of job security), they could prevent actors from forming a collective identity that carried authority at the bargaining table. But when producers could not control the *terms* of crisis, they lost some of their ability to control and define the terms of actors' subject identity.

Such was the case in 1933 when President Roosevelt charged industries, Hollywood among them, with the task of developing a "Code of Fair Competition." As part of Roosevelt's National Recovery Administration (NRA), the Code was to foster labor–management cooperation while also guaranteeing labor better working conditions and the right to organize and bargain collectively through representatives of their own choosing. Producers were able to obtain Code provisions that protected certain monopolistic practices such as block booking, but they could not prevent the federal government from interfering with and dictating the terms of labor–management relations. This breakdown of producer control shifted the terms of crisis and gave talent groups an opportunity to assert their labor power. But before screen actors could define a collective identity and a strategy of discursive self-representation that was effective in labor negotiation, they were forced to reevaluate their previous identificatory labels and affiliations. Their task was complicated by the fragmented and powerless state of actors' unions as well as the opposition of studio executives. It was only by establishing a group identity around the notion of collective labor through the Screen Actors Guild that actors finally were able to emerge as a powerful bargaining unit.

Institutions of Labor

According to screenwriter Frank Woods, one of the organizers of the Academy of Motion Picture Arts and Sciences, the Academy came into existence at a time when

> it seemed like a heaven born instrument for the welfare of the entire industry. Each of the talent classes had grievances with no medium for their adjustment. But more than this and of greater importance, as some of us viewed it, the screen and all its people were under a great and alarming cloud of public censure and contempt.[7]

Actors' Equity would have been a viable medium for actors' grievances if producers had recognized the union as the official bargaining representative of actors and if unionism had been a popular idea among high-ranking screen actors at that time. In lieu of this, the Academy at least allowed actors' voices to be heard. While actors were aware that the producers' branch of the Academy could exert an influence on issues affecting actors' working conditions, "they also hoped that friendly contact [with producers] would accomplish more good in the long run than would militant conflict."[8]

In addition, the Academy was not to function solely as an arbitration unit for labor–management negotiations. According to its charter, the Academy was organized for the

1) settlement of disputes within the industry by conciliation and arbitration,
2) protection from outside assault,
3) improvement in character and quality of production, and
4) promotion of the good repute and standing of the screen and its people within the industry and with the public.[9]

According to Woods, this last purpose was the one around which "all could unite." Due to the furor over film morality and Hollywood stars' immorality throughout the early part of the 1920s, prominent members of the industry felt that "some constructive action seemed imperative to halt the attacks and establish the industry in the public mind as a respectable, legitimite institution, and its people as reputable individuals."[10] It was this goal, and not the desire to head off unionism, says Woods, that provided the primary motivation for the Academy's formation.

The Academy's primary goal may not have been the negotiation of labor–management disputes, but the organization nonetheless became the sole arbitration unit for actors' labor between 1929 and 1933. Although it had remained "neutral" during Equity's union drive in 1929 (see chapter 2), the Academy stepped in immediately after the union's defeat to establish a minimum basic contract that would determine the codes of practices governing producers' relations with actors.[11] Murray Ross argues that while the Academy kept actors from unionizing, it also gave them their first tastes of the fruits of collective negotiation. Between 1929 and 1933, "the Academy settled 344 major actor cases, effected more than 3,600 interviews and minor adjustments, and collected approximately \$112,000 for the actors—an average of more than \$500 a week."[12]

Members of Equity had a different interpretation. First of all, they claimed that many of the important concessions achieved through the "company union" were made in response to Equity's relentless insistence upon fair practices. In addition, they felt that these concessions did not ultimately benefit actors. As an *Equity* magazine editorial explained,

> Invariably, the Academy has decided that it did not want a fight, that it did not want to go as far as Equity . . . and that it was appointing a committee to confer with the producers to settle whatever difficulty of the moment might be. Some time later the committees would announce that an amicable settlement had been reached. Each time some of the concessions demanded had been granted, though, it was almost never in full, and the standard contract, or code, or whatnot generally left the producers in full control of the situation.[13]

In contrast to Equity's desire for collective representation, the Academy permitted actors' grievances to be heard only on a case-by-case basis, and settlements were reviewed by a group of eminent screen actors (through the Actors' Adjustment Committee) or by a single representative from the Academy's actors' branch.[14] Through these policies, the Academy structurally reasserted the political hierarchy among actors while ensuring the labor power differences between actors and producers.

Relations between actors and the Academy worsened as the effects of the Depression hit the film industry. Between 1929 and 1933, the studios had made enormous investments in real estate (mostly theaters) and subsidiary industries (such as music publishing and recording). With a shrinking box office and with available capital tied up, studio heads began tightening their control over production costs, and "all studio employees, from stars to extras, felt the pinch of rapidly shrinking opportunities for employment and decreased earnings."[15] Studios began to pressure actors into taking voluntary cuts in the name of economic conservation. At Warners, for example, Darryl Zanuck persuaded Jimmy Cagney to take a $500-per-week salary slice, and Ruth Chatterton agreed to do an extra picture without pay.[16] Studios also began to violate the codes of practice established by the Academy. Terms for freelance actors were slashed as much as 50 percent, stunt players no longer received compensation for injuries, and conditions for extras deteriorated. By 1933, studios were boldly instituting salary cuts across the board. Although these cuts flagrantly violated the terms of actors' contracts, the Academy did not intervene. And, according to some observers in Hollywood, morale among workers had plummeted to such a point that they felt it useless to voice their grievances to the Academy.[17]

The situation erupted in March 1933 when President Roosevelt declared a nationwide bank moratorium. With capital frozen, the studios proclaimed a national emergency and confronted the possibility of an industry-wide shutdown. An Emergency Committee of the Academy, which was formed to review the crisis, sought to forestall such a possibility by proposing an eight-week salary waiver (to be in effect from March 4 to April 30). Using the familiar Depression-era adage that "half a loaf is better than none," the Committee asserted that this plan was the only way to prevent the studios from closing completely.[18] In accordance with the first point of the four-teen-point plan, all studios shut down production on the day of the announcement to allow the various branches of industry (as represented by the Academy) to hold mass meetings and vote on the proposal. According to a report in *Variety*, the "stormiest meeting" was held by actors who refused to vote on the salary waiver plan or to cooperate with the Academy board of directors. Other branches of the Academy responded in a similar fashion, and unionized craft guilds "unanimously voted to call out a strike unless the producers agreed to rescind their orders for a cut."[19] These responses were "precipitated by a studio belief that the Emergency Committee . . . had attempted to put something over on them."[20] Following these meetings, in which industry employees expressed such strong opposition and distrust, the trade and union presses widely speculated about the collapse of the Academy.

Though the Emergency Committee did not change the spirit of the plan, it did attempt to make the terms more equitable by creating a sliding scale. High-salaried employees (including executives, directors, as well as some writers and actors) assumed the brunt of the loss through a 50 percent reduction in pay, while those receiving $50 per week or less were exempted from the cuts. Salaried employees in the middle range lost 25 to 35 percent of their earnings; in addition, the Committee rendered several hundred decisions on individual cases. Overall, film payrolls were cut by 67 percent, having dropped from $156 million in 1931 to an estimated $50 million. (These figures include the wages saved from some ninety thousand employees who were cut from the payroll over the two-year period.)[21] The only assurance to remaining employees was a stipulation in the fourteen-point plan requiring the full resumption of salaries at the end of the eight-week period. In the meantime, in an effort to monitor the financial crisis more closely and to appease the opposing factions, the Academy demanded access to the studios' books to determine which companies might be able to resume full pay scales before the end of the waiver period.[22] The Emergency Committee hoped that these measures would guarantee the majority acceptance needed for the plan's implementation.

Actors recognized that the national emergency called for stringent reform, but they unequivocally opposed the principle of a "horizontal wage reduction." Any agreement that allowed producers to breech employment contracts left them with no assurances. Indeed, on the day of the salary waiver announcement, industry leaders stated that "there is little chance of any of the slashed salary percentages being returned until the country's box offices reflect a decided up movement."[23] Executives threatened that salaries would be "sacrificed to zero" before they would permit the industry to shut down. In line with this ultimatum, leaders adopted a "walk or stick" attitude. Any employee who disapproved of the salary waivers was free to walk out of a contract. According to the Eastern studio executives and bankers, "the more walkouts, the better," as that would be one of the few ways studios could cut material overhead.[24]

Crisis management within the studios had thus developed into an "austerity politics." Although this strategy is based on or arises out of economic hardship, it functions to legitimate policy through an ideological framework of "lack" and to diminish corporate accountability to the workers. Austerity politics, says Lichten, not only asserts the domination of labor by capital, but reverses the past gains of workers while enhancing corporate power.[25] According to a press release issued by the soundmen's union, "Studios took advantage of the bank situation which climaxed their financial difficulties to demand cuts; [they] also took advantage of panic among employees."[26] Thus, management held labor hostage in the crisis situation. According to a report in *Variety*, studio leaders began "planting the spirit that the industry owes nothing to the cut employee but that such employee owes the industry for still being in a job."[27]

Amidst this crisis, the Artist Managers' Association (AMA), the primary organization for actors' agents, attempted to intervene. Claiming that studios would not reinstate salaries to the 100 percent mark at the end of the emergency, the AMA argued that the only way screen players could ensure their entitlements was to unionize in affiliation with Actors' Equity and be accorded full support by the American Federation of Labor (AFL).[28] The agents had their own agenda as well. At the time of the salary waivers, agents had been negotiating unsuccessfully with producers to become an additional branch of the Academy; the threat of unionization appeared to be a means to facilitate this structural change. But while agents may have proposed the union idea as a means to enhance their own power, it also was pivotal in redefining actors' perceptions of their collective labor position. For the first time in screen acting history, a union between high-ranking and low-ranking actors—even one that required the consolidation of stage and screen—began to make sense to Hollywood's prominent actors. They realized that their contracts were no longer secure and that "their positions [were] as much in jeopardy as the smaller people" in the profession.[29]

The actors' impending unionization and a strike threat by studio craft workers prompted several studios to announce that they would resume full salaries, but Harry Warner refused to follow suit. In fact, Warners announced that it would extend the waiver an additional week.[30] The Academy stepped in to challenge Warners's violation of the fourteen-point plan, seeing this as an opportunity to regain the faith of disillusioned actors. But their "challenge" only revealed the extent to which the Academy functioned as a puppet organization of the MPPDA. Conrad Nagel, Academy president, made the error of seeking Will Hays's intervention in the matter, and thus exposed the Academy's inability to act independently of producers. Although the Academy board of directors immediately called for Nagel's resignation, they were unsuccessful in gaining Warners' cooperation.[31]

In retrospect, this situation was to be the Academy's final, critical test on behalf of labor. The once solid Academy had already begun to splinter because of its inability to represent the labor interests of employee members. A group of writers had officially withdrawn from the organization on April 5, 1933, to form the Screen Writers Guild (SWG), and the other employee branches of the Academy (actors and technicians) were strengthening their ranks along trade union lines. Witnessing the fragmentation of their company union, producers also threatened to walk out. Several producers stated unofficially that they would not deal with any collective talent group outside the Academy and would rather "move the industry out of Hollywood than countenance more unions or organizations seeking closed shop conditions."[32]

Those remaining loyal to the Academy immediately took steps to reorganize its structure and purpose. Although the board of directors thwarted "a move on the part of some elements in the Academy to throw out the producers and cement the organization into a purely employee organization,"[33] it seriously considered the possibility of including an agents' branch of the Academy should the producers voluntarily walk out.[34] In an effort to democratize the Academy, the board also embodied some "radical departures" in its amended constitution: (1) the refusal of an annual subsidy from producers; (2) the barring from Academy office any producer empowered to sign contracts; (3) a modification of membership requirements that would allow anyone in motion picture production to join; (4) a change in membership dues from a set amount to 1 percent of a member's salary (to encourage a wider membership and also make up for the subsidy loss from producers); (5) the compulsory arbitration of all contract and salary disputes; (6) an imposed limitation on the board of directors' power; and (7) the expulsion of any member (producer or employee) for violating constitutional by-laws and ethics.[35] With these reforms, the Academy felt confident that it had established an organization free from producer domination and acceptable to the employee branches seeking liberalization.

Despite its reformation, however, the Academy was not able to substantially increase its membership. Rosters submitted in mid-May showed a total of 958 members compared to 876 in December 1932. Within the actors' branch, stars and featured players numbered 200, and freelance and supporting actors totaled 100.[36] The majority of lower-ranking screen actors still belonged to Actors' Equity. Even though the agents' attempt to rejuvenate Equity had failed, and the union remained inactive in Hollywood, the Academy did not present an attractive alternative. The producers' branch had remained in the Academy, which meant that the organization still carried the taint of company unionism. Though subsidies from studio producers were now restricted to research use, and were no longer depended upon to finance the Academy's operation, the membership continued to consult with producers on policy matters and to include their interests in codes governing labor practices.

Some of the Academy's more questionable actions involved producer attempts to regulate and control the agenting business. Just after the salary waiver crisis period, for example, the producers tried to establish a centralized casting bureau (variably named the Central Artists Bureau and the Artists' Service Bureau) to be owned and operated cooperatively by the studios. Producers referred to the bureau as "a common pool for all actors, writers and directors." But, according to an *Equity* report, "The scheme really was intended to lower salaries and prices for screen material and contemplated the abolition of competitive bidding for creative talent."[37] Seeing the casting bureau as "an insidious device" designed by producers to control "both sides of the bargain," a group of actors, writers, directors, and agents combined their efforts to defeat the proposal.[38] (Though I was not able to substantiate it through research, another reason that the talent groups may have opposed the bureau is that it appeared too similar to the Central Casting Bureau, the job clearinghouse for screen extras. Higher-ranking talent, in other words, would not have wanted the same type of treatment by the studios.) Shortly after the artist bureau attempt, however, producers developed an artist-agent-producer code designed to eradicate (or at least minimize) "cases of racketeering, double dealing, arrogance, failure to live up to obligations, semi-legal trickery and the feuds between producers and agents which have caused loss to artist, agent and producer alike."[39] This code, which was endorsed by a majority of the Academy membership, obviated the need for an agents' branch of the Academy while ensuring producer control over the agenting business.

The Academy also conducted investigations into actors' working conditions with the intent of improving the actor's lot. But these actions often backfired. For example, as the actors' branch of the Academy was preparing a code of ethics that would guide investigations into the working conditions of contract and freelance players, branch chair Adolphe Menjou announced that actors would "confer with producers before arranging the code."[40] For actors who had become dissatisfied with producer control over conditions and policies in the Academy, this particular incident proved to be the final straw.

In late June of 1933, several ex-Academy members joined forces with a group of freelance players and Equity members and decided to form a new legal entity: the Screen Actors Guild (SAG). On July 12, two days after Menjou's announcement, the new organization went public. Since its eighteen founding members included several leading actors (Alan Mowbray, Ralph Morgan, C. Aubrey Smith, James Gleason, and Boris Karloff), the organization believed it could draw together the industry's high-ranking and low-ranking screen players. The founding members also felt confident that the SAG would provide a viable alternative to existing organizations (the Academy and Equity) since the Screen Actors Guild was neither a company union nor a theater union, but a union of and for motion picture actors. Unlike the Academy, the newly formed guild promised "eventual closed shop conditions, and an all-employee body that [would] be as strong in pictures as Equity [was] in legit."[41] David Prindle notes, however, that "the choice of 'guild' as opposed to 'union' was no whim." Many of the guild's organizers, "despite their activism, were socially and politically conservative, and would have been uncomfortable in any group that featured 'union' in its title." Thus, while the actors' group was certainly union-minded, it followed the example of the Screen Writers Guild and chose a name that "harked back to medieval associations of artisans."[42]

The formation of the Screen Actors Guild had been actively encouraged by the Screen Writers Guild, which was attempting to build a unified block among talent groups. The SWG felt that if the actors' guild was accepted, the directors would be next in line to unionize. Their ultimate goal was to establish a three-way alliance, plus a tie-in with agent members of the AMA, to achieve a pact that would have "the agents representing none but guild members, and none of the latter doing business except through AMA affiliates."[43] This alliance would be much more favorable to labor than the artist-agent-producer code developed by the Academy. In general, by breaking free of producer domination, the talent guilds would have greater control over defining their professional identities and working conditions.

Writers and actors were also reacting against the Academy's involvement in the National Industrial Recovery Act (NIRA). Signed by Franklin D. Roosevelt on June 16, 1933, the NIRA was one of the programs established by the president's National Recovery Administration (NRA) to stimulate economic growth and promote interindustry cooperation. To this end, both management and labor were provided with government assistance and protection. Industry was sheltered from antitrust laws (which protected monopolistic practices such as price fixing and block booking), and labor was guaranteed minimum wage and maximum hours as well as the right to organize and bargain collectively through representatives of their own choosing. The major difficulty posed by the NIRA, however, was that industry personnel were themselves given the task of drawing up a "Code of Fair Competition" to meet NIRA specifications.

The NRA Code

According to *Variety* reports, Roosevelt's NRA directive made Hollywood "code crazy." Since only one code would be accepted in Washington, D.C., on "Code Day" in September, various industry groups hurried to develop their particular versions of Fair Competition. By late June, General Hugh S. Johnson, chief administrator of the NRA, had received nearly a hundred petitions from studio personnel.[44] Members of the MPPDA, however, immediately established their authority to represent studio interests. While claiming to represent the majority, the Hays group vowed to "defeat any insurgent factions" and merge all codes into one complete formula. The producers were clearly more interested in the monopolistic provisions of the NIRA than they were in Section 7a, the portion that dealt with the rights of labor. The matter of minimum wages and maximum hours, for instance, had "not been discussed at any of the regulation conferences" during the first two weeks of consultation.[45]

Under the direction of J. Theodore Reed, the president succeeding Conrad Nagel, the Academy reasserted its commitment to interindustry cooperation and claimed the right to represent all talent groups in the NRA Code. In late July, the Hays group finally submitted that "all branches of the producing end of the industry [would] have an opportunity to offer suggestions for inclusion" in the Code, and they agreed to appoint a committee of five, one from each branch of the Academy, to work on Code development.[46] Sol Rosenblatt, NRA deputy administrator in charge of the Amusement Division, thereupon appointed Reed to the official Code committee. Reed was to play a somewhat duplicitous role. Although designated as an "employers' representative" in order to assure his acceptance by the Hays group, he also oversaw the Academy's right to represent employees in NRA affairs.[47]

The question of who qualified as an employee and thus deserved protection under the NRA Code was posed by the screen extra. It was estimated that by 1933 some 200,000 claimed extra status. According to the studios, however, most of these so-called extras were drifters, who, when they got hungry or into trouble, described themselves as extras. Because extras had become an embarrassment to the film industry, the Academy vowed to "separate the wheat from the chaff and make the title of 'extra' as tough to get as a policeman's badge."[48] The Academy Code Committee ordered an investigation into the Central Casting Bureau and found that only 18,000 extras were officially registered with the Bureau and that, on the average, each screen extra worked only eleven and a half days per year.[49] Part of the problem was caused by "coast bums" who were willing to work for less, and had put experienced bit players out of work and driven down the average wage to $1.25 per day.[50] The Academy also found that favoritism within the

Bureau had become quite common, and that special requests for particular individuals made up 30 to 80 percent of the studio lists submitted.[51] The Academy's Standing Committee on Extras pledged to remedy these situations by representing screen actors at the NRA Code hearings and by instituting Code provisions that would set minimum wages, place a limit on the number of extras, and create a more equitable distribution of work.

Actors' organizations outside the Academy were suspicious of the Academy's sudden interest in and commitment to the screen extra. Ross, for example, argues that since the salary waiver period, many actors (especially those who still belonged to Actors' Equity) felt that the Academy's reformations were only "surface inducements to keep the artist groups in the Academy so that the latter could qualify as the collective bargaining agent under the NIRA."[52] The fact that the Academy had been invited by producers to participate in the Code's development also raised suspicions. The Code Committee's handling of the extra problem was read as merely an attempt to improve the studios' public image. In addition, given the labor power differences between extras and Academy members, the Committee's concern was perceived as a form of paternalism. In an effort to place control in the hands of extras themselves, AFL organizers outside of Hollywood attempted to unionize extras under the AFL banner, but their effort failed.[53] The newly formed Screen Actors Guild also lacked the financial and political strength to challenge the Academy. But when the completed NRA Code was released on August 28, the actors' suspicions about the Academy were verified, and alternatives to the company union's plans became possible.

In drafting the NRA Code for the film industry, studio heads had sought to increase their industrial authority and to curtail wages even more. They blamed their financial difficulties on high-priced talent and saw the national recovery program as a way to solve the problems of the star system that had developed in the studios. As Larry Ceplair and Steven Englund explain, the NRA Code "seemed to provide management with the opportunity to do its patriotic duty by reducing salaries across the board and getting away with it."[54] In an effort to ensure a systematic and federally sanctioned method of controlling actors, producers included clauses in the NRA Code designed to eliminate star raiding (Article VIII), to curb the activities of agents (Article IX), and to limit the salaries of creative artists (Article X). These clauses would become sites of bitter labor–management struggle in the months to come.

Producers already adhered to a practice of "disciplining" stars who were under contract by lending them out to other studios to appear in inferior pictures. While this practice occasionally played to the stars' advantage (e.g., when a low-budget production became a surprise hit), it usually impeded their careers; producers then pocketed the fees charged to the borrower (which were in excess of the contract players' specified salaries). Article VIII was to formalize this practice and permit producers to "loan" players without either their consent or the benefit of salary increases.[55] Thus far, star raiding had kept this practice in check. Because producers found it more profitable to hire talent that had already been developed, they often raided the talent of competitors. For stars, this system remained one of the few recourses available for obtaining better working conditions; an actor either accepted a better offer from another studio and broke the terms of the existing contract, or, with the threat of departure, was able to negotiate a new contract with the current employer. By 1933 star raiding was becoming costly to the studios. Producers maintained that because they developed and publicized stars at great expense they were entitled to keep them and net a fair return on their investment.[56] To ensure the elimination of star raiding, producers included a separate clause under Article VIII that would place a ceiling on wages, prohibit competitive bidding, and blacklist any artist who engaged in salary bidding practices.

Producers also sought to restrict the power of agents by calling for the inclusion of the Academy's artist-agent-producer code in the NRA provisions. Although agents were indispensable in managing business affairs, negotiating contracts, and supervising the professional career of the actor, producers often felt themselves forced into packaged deals that were costly and unsatisfactory. In order to obtain the services of a highly desired star, for example, the producer might also be manipulated by an agent into signing a written contract for a lesser star. Although stars could lose out on a packaged deal, or suffer consequences when their agent created a feud with a particular studio, such arrangements generally benefited actors (and thus their agents) while depriving producers of the power to develop their stable of actors without interference.[57] Most sectors of Hollywood (including agents) agreed that the agenting business contained unethical practitioners, and attempts had been made both inside and outside the profession to create a means of (self-) regulation or a standardized code of ethics, yet the producer-supported artist-agent-producer code obviously weighed in favor of studio management.

To further increase their own interests, producers attached additional clauses to Article IX of the Code. One would require agents to be *licensed by producers* and prohibit studio negotiations with agents "not parties to the code." Another would bar agents "from being present in producer interviews with their clients except on strictly financial matters."[58] This latter provision meant that an agent was forbidden to come to the aid of a client when other aspects of a contract came under dispute. Since the actor's only other source of contract protection was the producer-controlled Academy, actors were strongly opposed to provisions that would weaken or limit the agent's services. According to Actors' Equity, Article IX was actually "an attempt to legislate the agents out of the business."[59] Members of the Screen Actors Guild agreed. The producers, they said, were not really interested in a code of fair practice, but in making agents useless to clients: "An artist by himself is easier to deal with. . . . Take away his business aid, and he is easy pickings for the astuteness of producers."[60]

What was referred to as the salary-fixing clause was actually part of a provision governing contract negotiation. According to Article X, offers made to players under contract were to be made three to six months (depending on the actor's rank and salary) preceding the expiration of the contract. The studio under which the player was contracted would then have a "reasonable opportunity" to make a competitive offer. This unspecified allotment of time, however, caused some to fear that producers would also take up to three to six months to make a competitive bid. Actors felt that this "reasonable opportunity" would allow time for gentlemen's agreements to be made with the result that "salaries would be kept under tight control and at levels considerably below those to which they might rise untethered."[61] Meanwhile, producers were hoping that NRA officials would cooperate in providing a specific ceiling on wages within the Code.

The completed draft of the NRA Code of Fair Competition was thirty-some pages of legalese that addressed every aspect of the industry from local clearance and zoning boards to child labor. Although Academy members were in general accord with the NRA Code draft submitted by industry executives, they took exception to the clauses regarding agents' activities, star raiding, and salary limitations. The Academy did not oppose the agent-licensing clause, but it regarded the stipulation that denied an employee the right to an agent's representation as "going beyond the Academy code of fair practice" and asked for its deletion. The Academy sought modifications in the star-raiding clause that would reduce its applicability to a very small group and only to "expiring contracts which the employer did not have the right to extend by the exercise of an option."[62] On the issue of stars' salaries, the Academy vigorously opposed any wage-fixing clause and recommended that the producers' "reasonable opportunity" to respond to contract offers be limited to three months.[63]

The producers' draft of the NRA Code was sent to Washington and circulated among members of the Hollywood community in anticipation of the public Code debates. On September 12 and 13 (the two days set aside for debate in the nation's capital), those in the motion picture industry who wished to call for modifications in the document or to oppose any of its provisions would have an opportunity to speak. Actors were angered by the text of the Code and publicly challenged the Academy's right to represent all actors in the Code proceedings. They felt that the drafted Code indicated the Academy's acquiescence to producers and inability to protect labor's interests.

The Screen Actors Guild did not challenge the Academy in this forum because it did not have the necessary representation to make an appearance at the Code hearings, but several other actors' groups made their voices heard. Indeed, the U.S. Chamber of Commerce building (where the debates were held) became a dramatic arena of struggle where the various factions of the acting community asserted their identities as political subjects and their desires for self-representation. Hollywood extras, for example, rejected any association with the Academy and were represented at the Code hearings by a Washington lawyer. (Extras originally asked Mary Pickford, a leading advocate of extras' rights, to represent them at the Code hearings, but she delegated the task to the lawyer.) Although most of the recommendations made by the Academy's Standing Committee on Extras had been incorporated into the NRA Code, extras announced that they were writing their own code to provide for the inclusion of an arbitration board.[64] A group of 1,750 unaffiliated actors and directors, represented by Max D. Steuer, also voiced its resentment toward the Academy's assumption of authority over the entire creative wing of the industry. The group requested a ban on the lending of creative artists and condemned the clauses of the Code that prohibited star raiding and permitted producer licensing of agents.[65]

Though Actors' Equity had been forced into the background for several years, the union now stepped forward to assert itself. Frank Gillmore, president of Equity, pointed out that although the Academy had only 296 actor members, "54 per cent [1,418] of the actors in Hollywood getting screen credits [were] Equity members . . . [and] 80 per cent of character and small part actors whose names do not appear in credit lists also were Equity members."[66] Given the Equity majority, union officials asserted their right to represent all screen actors. They were also angry that Equity had not been consulted during the drafting of the NRA Code. Now, confronted with a completed document, the union could only submit arguments for specific deletions, additions, or modifications without the benefit of interactive debate with the Code Committee.[67] Gillmore blamed the Academy for its failure to procure better provisions and, in a heated debate, charged that the Academy was still nothing more than "a company union for the producers." Under these circumstances, he concluded, "players cannot possibly get fair representation through the Academy."[68]

Academy president Reed denied the charges of company unionism. As a sign of its independence from producers, Reed cited the Academy's commitment to the improvement of wages and working conditions for screen extras and modifications of the agent-licensing and antiraiding provisions. But Reed's defense of the Academy was undermined by Lionel Atwill, who spoke at the hearings on behalf of the actors' branch of the Academy. Atwill revealed that the five branches of the Academy were *not* in accord among themselves; although the other branches had ratified the Academy's Code recommendations, the actors had voted unanimously "to pull all the teeth out of the proposed instrument."[69] They opposed the amendments to the artist-agent-producer clause and the star-raiding clause, and also called for articles forbidding the loan of contract players and prohibiting freelance actors from playing in more than one picture simultaneously. These latter practices, asserted the actors' branch, added to unemployment rather than stemming it as the NRA intended.[70] Once it became clear to Code officials that the Academy was suffering from internal dissension and that it did not, in fact, represent the numerical or political majority of actors in Hollywood, the Academy wielded no more influence than other independent groups at the Code hearings. And Article VII, which named the Academy as "the clearing house through which all the producer-employee relations were handled," was stricken from the Code.[71]

Overall, labor groups received a boost from the Code hearings. When NRA deputy administrator Rosenblatt opened the official Code hearings on September 12, the issue of labor received top priority, and delegates of some sixty different union groups were present to voice their interests. On the second day of the hearings, Rosenblatt began the session with an announcement that "the 'open shop' principle [was] to be ruled out of the film industry's code."[72] But when the second draft of the Code was released at the month's end, it contained the agent-licensing, salary control, and antiraiding provisions that actors had fought against. Knowing that Reed was a member of the committee that drafted the final provisions, actors who had remained

in the Academy finally became convinced that the organization was "a pro-
ducer-ruled body, misrepresentative of the acting profession," and twenty-six
of Hollywood's most prominent actors resigned from the Academy with the
idea of forming a new organization.[73] The group sent telegrams to Rosenblatt explaining their resignation and protesting the inclusion of Articles IX
and X in the NRA Code.

The Screen Actors Guild was able to convince the ex-Academy members to join forces with their group. Though the Code Committee was preparing a final version of the document to be submitted for official signatures,
the Guild felt that the influx of new actors (especially such prominent ones as
Adolphe Menjou, Robert Montgomery, Paul Muni, Eddie Cantor, Miriam
Hopkins, and the four Marx brothers) would give them the strength necessary to block the unwanted provisions from the NRA Code. The Guild
launched a determined publicity campaign, sending numerous telegrams to
officials in Washington and feeding the local press with "threatening stories
of what the creative artists might do if the obnoxious code provisions were
not peacefully deleted."[74] It also undertook a massive recruitment drive,
which culminated in a mass meeting at the El Capitan Theatre in Los Angeles on October 8. Of the eight hundred actors in attendance, five hundred
joined immediately, and the others agreed to submit membership forms the
following day. Through this event, the SAG was able to recruit the remaining
actors from the Academy, thus causing the final collapse of the company
union.[75] According to a Guild report:

> The Academy is a joke. There if ever was an opportunity to create
> a permanent tribunal of high ethics and fair play between producer
> and artist. And what happened? The producers used it as a pot in
> which to stew their little political potatoes. They polluted its high
> principles and used it for subversive purposes until, at a time of cri-
> sis, talent . . . having lost all faith and confidence in it, walked
> out—en masse.[76]

With the secession of the actors' branch, the largest branch and the
"backbone" of the Academy, the organization was left with few employee
members and lost its power as an arbitration unit. Some of the ex-Academy
members, however, immediately assumed a position of authority and leadership in the Screen Actors Guild. Eddie Cantor's inspiring speech, calling
for "an actors' organization 'of, by and for actors alone,' " won him the office of president, and other ex-Academy members were elected to the remaining seats: Adolphe Menjou, Fredric March, and Ann Harding became

vice presidents; Groucho Marx, treasurer; and Lucille Gleason, assistant treasurer.[77] Their first action was to compose a strongly worded telegram to President Roosevelt protesting the salary control provisions of the NRA Code and charging producers with extravagance and financial mismanagement. The Guild asserted that "the motion picture companies [had] not been bankrupt by salaries to talent, but by the purchase and leasing of theaters at exorbitant prices," and actors were concerned that the NRA Code was giving "oppressive powers to the very people whose mismanagement [had] brought this industry to the verge of bankruptcy."[78] The Guild also reminded Roosevelt that the salary-fixing clause would fix maximum rates of pay, something that the NRA had expressly forbidden:

> This is so un-American in conception that it makes one shudder to think it is being seriously considered in Washington. . . . Every manual worker in every industry is threatened to have a rate of pay fixed for all time beyond which he cannot progress.[79]

The telegram ended by asking the president to make a thorough investigation of the provisions under debate.

According to unofficial, though undisputed reports, Roosevelt had been "an acute observer" of the film industry's code and had reached some definite conclusions about the industry and its operations. Among these was the opinion that " 'excessive' motion picture salaries must be reduced for the common good."[80] He apparently was alarmed when he learned that various individuals in Hollywood received compensation "several times greater than the $75,000 salary of the nation's Chief Executive."[81] He had thus ordered NRA officials to compile a report on the salaries of directors and stars, especially child stars, to see if they were earning more than was "conscionable."[82] The initial reports that "466 individuals . . . drew 51 per cent of the producers' payroll"[83] indeed seemed to confirm the administration's concerns. These reports, which were released only a couple of weeks after the Screen Actors Guild's telegram, did not address the legal aspects of salaries, nor did they appear to address any of the Guild's concerns.

Several days later, upon hearing rumors that NRA officials would establish a salary-fixing board in the Code, the SAG threatened a walkout. Actors expressed their unequivocal opposition to the incorporation of any salary limitation (including those provisions that would affect salaries through star raiding and agent representation). Speaking on behalf of the SAG, Guild president Cantor issued a warning:

If proposed Articles 9 and 10 are included in the motion picture code, not an actor of importance will work in a studio which signs the code. These provisions are un-American. The days of slavery are over![84]

Cantor also stressed how few actors received "headline" pay (fewer than fifty received between $3,000 and $10,000 a week). "The actor always has been the target for attack when it comes to salary," he stated, "[but] you don't find Jackie Coopers and Baby Leroys every day." Noting that "the big fellows in the guild are not concerned about salaries," Cantor pledged that the Guild was "out to protect the little fellow that the [Roosevelt] administration should protect."[85] After this speech, scores of low-salaried actors joined the Guild, swelling its ranks to more than a thousand members.

Producers had initially supported the public disclosure of stars' high salaries because they felt it would help them to secure a salary-fixing board in the Code. In late September, for example, producers reportedly asked Rosenblatt to establish a board to investigate (or disclose) actors' salaries. But producers otherwise maintained low visibility on the matter. Their rhetorical strategy was to absolve themselves of all responsibility. By admitting that "Hollywood [was] beyond its control," industry leadership claimed that "only the government [could] call a halt to the ascent of salaries."[86] This strategy began to work against them, however, once the Screen Actors Guild became a powerful force in the Code debates and began to publicize the high salaries of executives as well as the low (or nonexistent) earnings of bit players and extras. Through its spokesman, Eddie Cantor, the SAG was relentless in voicing objections, presenting data, and pressuring Rosenblatt into listening to Guild demands. In an effort to "appease the government" (or possibly the new actors' union), industry leaders thereafter instructed their press agents "to play down salaries."[87] One reporter suggested that "the film industry was nervous with fear that a Government inquiry into salaries might throw the spotlight on golden secrets hidden during the depression."[88]

Government officials also began to back off from their commitment to limit salaries. Legal experts warned that any government attempt to fix artists' salaries would indeed set a dangerous precedent.[89] When the final version of the Code was released, the salary-fixing clause was thus reworded as an ambiguous provision that deemed "excessive" salaries an unfair trade practice. Under this clause, the Code Authority (the government body charged with enforcing NIRA regulations) could fine producers $10,000 for the payment of salaries "unreasonably in excess of the fair value of personal services," though producers faced with fines were under no obligation to *change* the salaries.[90] Articles IX and X were also "softened" to allow actors a better bargaining position in matters pertaining to employment and salary.

The Code was finally forwarded to President Roosevelt, who agreed to sign it into law by the end of his Thanksgiving holiday. As a final measure, however, the president invited producer Joseph Schenck and SAG president Eddie Cantor to his home in Warm Springs, Georgia, for "social visits."[91] Cantor described his meeting in favorable terms:

> I found the President warmly sympathetic to our problems, a delightful gentleman, and a real friend of every man who works for a living, be he large or small. I left with the President a brief setting forth our position in regard to those iniquitous clauses. The President promised to read it.[92]

When Roosevelt signed the Code a few days later (November 27, 1933), he sided with labor. The antiraiding, agent control, and salary curtailment clauses were suspended, and seven months later the suspensions were made permanent. Roosevelt also made formal acknowledgment of the SAG's right to represent actors by appointing two Guild members (Eddie Cantor and Marie Dressler) to the Code Authority, the body charged with overseeing the Code of Fair Competition in everyday practice.[93]

The Positions of Labor

In the construction of history arises the problem of narrative and, with it, the tendency toward unself-conscious discourse, seamless causality, and teleological bias.[94] On the other hand, a history without narrative is a history without subjects. As Teresa de Lauretis has pointed out, "Subjectivity is engaged in the cogs of narrative . . . the very work of narrative is the engagement of the subject in certain positionalities of meaning and desire."[95] My own desire concerning the narrative of actors' labor involves fantasies of closure and achievement, of seeing actors as the victorious subjects of a self-determined history. But this desire, no matter how much it may have seeped into the prose of my preceding account, and no matter how seductive or politically pleasurable it may be, cannot be fulfilled. The narratives of actors' labor lead to no easy conclusions, but rather constitute a complex set of social relations and social practices involved in a shifting process of struggle over power.

Although actors managed to block the ratification of the producers'

Code provisions, their victory was tempered by ongoing struggles with producers. In their crusade against eliminating certain clauses in the Code, for example, they neglected (or had no time) to include substitute provisions governing actor–producer relations.[96] The producers, meanwhile, claimed their own victory with the final version of the Code. According to one reporter, the lengthy and elaborate document signed by NRA officials was little more than "a charter for self-regulation."[97] Having achieved protection from antitrust laws, producers remained unwilling to uphold the laws pertaining to labor. Though the NRA maintained a closed shop for the industry, producers continued to conduct their affairs with labor according to traditional open-shop policies and refused to recognize the unions as formal bargaining units.

As a writer for the *Motion Picture Herald* predicted, nothing basic would be changed in the structure of the film industry with the passage of the NRA Code: "The net result . . . will be that motion pictures will be made by the same persons, sold by the same persons and consumed by the same people, and on very much the same terms."[98] The effect of the NRA period was a solidification of the industry's oligopolistic structure and division of labor that had already developed by 1933. The NRA furthermore established a parallel development of "big management" and "big unionism" that locked into place the imbalance of power between producers and employees, thus assuring that labor power relations between actors and producers would remain relatively unchanged until the breakup of the studio star system in the late 1940s and early 1950s. In the aftermath of the NRA, after repeated attempts to negotiate a new set of basic minimum contracts with producers, actors reached the following conclusion:

> There is apparently no penalty for failure for a motion picture executive. The same group of men who have taken millions of dollars out of the American public through their manipulation of the motion picture business are still in control. With few exceptions they have never contributed anything . . . to the advance of the art. Yet these same men arrogate to themselves a despotic feudalism over the working conditions of those who actually make pictures. . . . The actors have exhausted every effort to agree with the producers on working conditions. They have been tricked, hamstrung, and lied to. Every dishonest practice known to an industry the code of ethics of which is the lowest of all industries has been resorted to by the producers against actors.[99]

It was not until the Basic Minimum Contract of 1937 that actors achieved better working conditions. In this agreement, "minimum rates were set, continuous employment was guaranteed . . . [and] an actor's right to a 12-hour rest period between calls was established."[100] The contract also established procedures for the arbitration of Guild–producer disputes.

The struggle between actors and producers that occurred during the period of the Screen Actors Guild's development, however, was not so much a war of policy as a war of positionality. What was important for actors, as subjects, was their ability to define their own positions within the studio system and claim their right to self-representation. Some of the lingering conflicts between stage and screen actors were resolved in November 1934 when Actors' Equity and the Screen Actors Guild entered into a "contract of affiliation," which sorted out issues of jurisdiction, governance, and membership. Equity agreed to relinquish its authority over the motion picture industry pending the Guild's official recognition from the AFL chapter, the Associated Actors and Artistes of America. The contract further specified the conditions under which an actor would be required to hold joint membership, and the means by which that member would pay dues or be suspended. Recognizing that it was "the purpose of both organizations to further the interests of actors, wherever they work," the contract also included a bylaw that pledged solidarity in the event of a strike by either party.[101]

Because the Guild's three thousand members included both high-ranking and low-ranking actors, the union was still left with some internal divisions, and a number of adjustments were necessary. As Murray Ross points out, the SAG initially had no intention of taking up the extras' cause. When the union sought to increase its membership in early October 1933, it turned to prominent actors disillusioned with the Academy, and the rallying issue became the protection of high salaries. But when hundreds of extras showed up at the mass meeting at the El Capitan Theatre and "implored the guild to keep them as part of the organization," the SAG could not refuse their request.[102] The agreement to represent extras fortunately worked to the Guild's advantage during the NRA Code debates. Aside from any solid commitment the prominent Guild members may or may not have had toward the screen extra, references to the "little fellow" served as an effective rhetorical device that shifted emphasis away from the issue of exorbitant salaries while simultaneously masking labor power differences among actors, and allowing them to present a powerful, united front.

In the period following the Code's ratification, the Guild maintained this rhetorical strategy of positioning, referring to itself as "the economic fraternity of the star, the freelance player and the extra."[103] But the union was clearly "in the uncomfortable position of relying on its stars to lend support to a collective-bargaining process that mainly benefit[ed] its rank-and-file members."[104] The actors' victory in eliminating the NRA salary-fixing clause did not necessarily enhance the earning power of stars. A new structure of wage practices had been developing over the preceding years, and, by 1933, most headline stars were paid by the picture and were engaging in profit sharing of the picture's revenue. This arrangement effectively placed a ceiling on wages such that a star could only command a six-figure salary at the peak of popularity.[105] But even though the highest-priced stars were subject to limitations and control by the studios, they still enjoyed a degree of bargaining power that was inaccessible to others.

On the other hand, in a report submitted to NRA deputy administrator Rosenblatt, the Screen Actors Guild concluded that a majority of Hollywood's population was "barely able to keep alive on the scraps they get from the industry's table." An estimated 71 percent of the actors who worked in 1933 earned less than $5,000; only 12 percent earned between $5,000 and $10,000; and the high-priced category of stars (those earning more than $50,000 per year) constituted only 4 percent of the total acting profession.[106] After the Code went into effect, the Guild discovered the extent to which it would need to work on behalf of its lower-ranking members as it took over many of the functions that previously had been the province of individual studios or the Academy. This included, for example, the surveillance of working conditions for screen extras. Upon achieving the Basic Minimum Contract of 1937, a 100 percent Guild shop was established for all extras. To ensure studio adherence to this policy, however, it was up to the Guild to police production lots for nonunion labor. The Guild also worked closely with the Central Casting Bureau, reclassifying the various skill areas and compiling exhaustive files for each of its registered players.[107] According to Ross, producers "found it a great convenience to shift the responsibility for extras to the guild."[108]

As much as the extras may have benefited from their membership in the Screen Actors Guild, they remained in a relatively powerless position within the union structure. Following the NRA period, screen extras were given relative autonomy through the establishment of a junior guild. The extras governed themselves through this branch with their own council and officers, but they were subject to the veto power of the SAG's board of directors. The relationship between high-ranking and low-ranking actors that this structure established was not substantially different from the relations maintained by the actors' branch of the Academy. While taking into consideration the different needs of the two ranks, this structure assumed a hierarchization of the acting profession that perpetuated paternalism. The extras had no direct voice in the senior organization, which was presided over by ex-Academy members. Several SAG officers had served on the Academy's Special Committee on Extras, and since the NRA Code had incorporated practically all of that committee's recommendations, the Guild was in effect functioning as the Academy in absentia.

The chief difference between the SAG and the Academy was, of course, the absence of producer members. But once the Guild achieved the Basic Minimum Contract of 1937, it adopted the Academy's former stance of interindustry cooperation. The successful negotiation, stated the Guild, "is not a conquest of the producers by the actor; it is the triumph of an ideal—a victory for the entire industry."[109] Ceplair and Englund argue that, thereafter, the SAG became complacent, turning into "the most tame and domesticated of unions" in Hollywood.[110] The Guild refrained from troublemaking in an effort to convince producers that the fair and responsible treatment of actors was in everyone's best interest. When conflicts did arise, the Guild frequently consulted the interpretations and decisions made by the Academy during its pre-1933 period of arbitration in actor–producer relations.[111]

From the standpoint of labor ideology, it appears that film actors "sold out" or were recuperated into an institutional structure dominated and determined by producers. But, as Ross points out, "The Marxist doctrine which attempts to explain unionization as a function of 'proletarianism' throws no light on this situation."[112] The actors' eventual acceptance of themselves as workers never translated into class solidarity or "class consciousness." As Eddie Cantor explained in an editorial for the inaugural issue of *The Screen Player* (March 1934), "It is not a matter of class consciousness or class struggle, but sober realization of the economic barrier between the employer and the employed."[113] But although actors' labor struggles in the 1930s did not end with their radical politicization (as class subjects), the fact remains that the actors themselves felt they had formed an organization "of, by and for actors alone." Actors drew the lines of demarcation for actor–producer relations even if they did not significantly alter labor power differences. By defining themselves collectively as workers, actors had rebelled against the paternalism of studio executives and placed their loyalty with each other and with the acting profession itself. Union membership thereafter signified an actor's support of other actors, regardless of rank or studio affiliation.

Actors' struggles over labor and subjectivity must be viewed within a broader context, however. The fact that producers still controlled and defined actors within the sphere of exchange, and encouraged the nation's audiences to identify with the object status of actors' commodification, constantly undermined actors' efforts to control the terms of their (self-) representation. Further investigation into actors' subjectivity thus reveals the extent to which actors were caught between production and consumption, between the material conditions of labor and the discursive representation of that labor. The historical analysis in this chapter has attempted to examine actor–producer conflict in the sphere of labor exchange and to privilege the voices of actors as workers; the following chapter analyzes how actors' voices were silenced within discursive representations of labor controlled by producers.

4 / Discourses of Entertainment

The film industry depends upon the production of entertainment for profit. While entertainment is ostensibly produced and sold in the marketplace in film form, the long-term effects of capital accumulation hinge on packaging the image of entertainment itself. In other words, the cumulative wealth of the American film industry can be attributed not only to box office receipts, but also to the successful promotion of a discourse about entertainment. Hollywood assumes a central role in the realization of this discourse. As the entertainment "capital," Hollywood is both the site of economic production and the source or repository of ideological (re)production. The major film studios have thus sought to foreground particular aspects of the production–consumption process, and to purposefully create or hide other aspects, in order to control an image of the industry and of entertainment that influences audiences' reception of films as well as their perception of Hollywood.

When the "ideological capital" of Hollywood is destabilized by economic or political events, the studios must find ways to prevent the dominant discourse of entertainment from being undermined. In 1933, for example, when the banking crisis occurred, and labor–management conflicts arose over the development of the industry's Code of Fair Competition, the image of Hollywood as a source of entertainment and wealth was threatened. But by rewriting the events of the National Recovery Administration (NRA) period to reaffirm their discourse of entertainment, industry leaders were able to maintain a "profitable" public image. This revisionist strategy was made possible by diverting attention away from the industry's internal struggles and asserting its commitment to loftier ideals. More specifically, conflicts involving trade and labor practices were minimized in public discourse by subsuming them within broader industrial policies that supported national goals and dominant social values.

What is significant here is the way in which the major studios relied upon the ideological power of narrative to win public and government support for their business practices. Though they clung to certain trade and labor advantages, this alone was not enough to ensure their profitability in the industry. The major studios needed external support—an ideological consensus—which could only be attained by capitalizing upon the dominant discourse of entertainment. As stated in the 1933 Report of the President to the Motion Picture Producers and Distributors of America (MPPDA), "The stability of the motion picture as an entertainment art does not rest upon bricks or mortar or upon money or men. It is dependent upon public appeal and public confidence."[1] Studio executives, who fancied themselves entrepreneurs of unusual courage and vision, commonly tried to win public confidence through a rhetoric of service. According to a statement in the Production Code Administration Annual Report of 1931:

> There should be a clearer understanding of the value of motion pictures in this period of depression, when they are bringing happiness and pleasure to millions of people, many of whom are laboring under the greatest distress in their homes and in their businesses. We are actually performing heroic essential work.[2]

The studios' self-proclaimed role as the producers and guardians of America's pleasure was complemented by a discourse of morality. For, as stated in the MPPDA Annual Report of 1932, "Entertainment possesses in itself and of itself a moral value and is a vital necessity to the millions whom we serve."[3]

Issues of morality intensified during the NRA period, in part because this period coincided with the years intervening between the adoption of the Production Code in 1930 and the enforcement of the Code in 1934. Studio executives discovered, moreover, that by constructing a discourse that presented the NRA in moral terms, and by infusing this discourse with references to patriotism and quality production, trade and labor practices could be rewritten more easily into a discourse of entertainment. Thus, the studios actively disseminated these discourses through fan magazines, the trade press, and even films themselves (see chapter 5) as a means to reinforce their political and economic dominance.

Studio Morality

Studio executives maintained their dominance most effectively under the direction of Will H. Hays and the policies of the MPPDA. The stated purpose of the MPPDA was "to foster the common interests of those engaged in the motion picture industry in the United States by establishing and maintaining the highest possible moral and artistic standards in motion picture production."[4] This combination of moral and managerial concerns (under the auspices of artistic standards) was strategically turned to its full advantage during the NRA period. Although the guidelines governing motion picture morality had been formally adopted in the Production Code of 1930, producers issued a "Reaffirmation of the Code to Govern the Making of Talking, Synchronized and Silent Motion Pictures" on March 7, 1933—the day President Roosevelt proclaimed the nationwide bank moratorium. The Hays group had called a meeting to review the industry's financial situation and concluded that the solution to the national emergency was an eight-week salary waiver (see chapter 3). But, as Hays describes in his memoirs, "out of that [same] meeting, as out of what the Bible calls 'the refiner's fire,' came the answer to those who were advocating retreat from [their] original purposes, and in particular from the forthright moral position of the Code."[5]

Adherence to the Production Code's list of "Don'ts and Be Carefuls" had slackened considerably in the Depression years between 1930 and 1933 as producers sought to increase box office receipts through "spicy" scripts filled with sex and gangsterism. According to Hays, this route had taken the motion picture industry "deeper into the uncharted sea of [moral] depression," and "at the height of the storm, the [MPPDA board of] directors even debated abandoning ship." Yet, when the industry was put to the test, Hays asserts that producers "held to their faith in the motion picture, I held to my faith in them, and we both held to our faith in the public."[6] It is noteworthy that Hays relies on metaphors of crisis and external calamity not only to exonerate the "aberrations . . . of an industry struggling with economic depression,"[7] but to explain why the MPPDA chose to reaffirm its commitment to the Production Code on this particular date.

The producers' seemingly sudden conversion on the eve of the bank holiday must be seen in context, however. For the MPPDA's claim to public responsibility cannot be separated from its desire for self-regulation. The Hays group held the conviction that "only through a larger measure of self-government could business escape the paralyzing hands of government bureaucracy and politics."[8] Fearing that the Roosevelt administration would extend federal control over media industries, the MPPDA attempted to avert state domination through a twofold strategy. First, producers admitted that the industry was out of control. Subsequently, they resolved to "discipline [themselves] for the mutual benefit of the industry and the public" by adhering to a strict code of film production.[9] As stated in the Reaffirmation of the Code,

> It is inevitable that during a period such as we now face disintegrating influences should threaten the standards of production, standards of quality, standards of business practice built up and maintained by cooperation. . . . Not only is a continuous supply of motion picture entertainment doubly essential in these times of confusion and distress, but the tendency toward confused thinking and slackening of standards everywhere re-emphasizes the importance of the progressively effective process of self-discipline by which the moral and artistic standards of motion picture production have been steadily raised during the past eleven years.[10]

The MPPDA's strategy duplicated that of the Roosevelt administration, which, in admitting economic crisis, justified the implementation of its national recovery program. In an era of austerity politics, this industrial discourse of crisis, confession, and moral conversion was effective in permitting the MPPDA to maintain control over business practices.

Throughout the NRA period the producers' commitment to film morality served as a pledge of cooperation and good faith that simultaneously diverted attention away from and legitimated the dominating practices of the MPPDA. The organization's control over independent producers, exhibitors, and labor groups was thus rendered invisible if not inherently justified on the basis of its moral concern for the economic future of the motion picture industry and of the nation. When Roosevelt signed the National Recovery Act into law in June of 1933, the MPPDA was quick to pledge its support while also making it clear that the regulation of a motion picture code under the NRA would pose "no new problem":

> We had been operating for three years under comprehensive production and advertising codes; our trade practices had been examined and commended by the Federal Trade Commission; our wages

and working conditions were sometimes referred to as "the best in the world."[11]

By rewriting its economic domination as a commitment to high standards of trade, the MPPDA hoped to protect itself—and succeeded—against government interference. Roosevelt himself assured Hays that the association's self-regulatory practices were "working as well as any plan could work."[12]

The MPPDA's duplicitous use of moral discourse was further reflected in its attitude toward the inclusion of a morality clause in the NRA Code. According to a report in *Variety*, the industry's "moral machine" created one of the "strangest paradoxes" in the NRA proceedings. On one hand, producers proposed the adoption of a clause that would showcase their commitment to film morality and indicate their general support of the NRA. On the other hand, major producers decided "not [to] incorporate the picture morals preached to women's clubs all over the world" because they did not wish to freeze a set of guidelines into a statute that invited public criticism.[13] The Hays group also feared that the inclusion of specific guidelines would permit the NRA Code Authority to impose its jurisdiction over the industry and enforce movie censorship. Thus, according to industry spokespersons, it was better "to be a little ambiguous and to remain in a liquid position, thereby conforming the moral issue from time to time to ever-changing public taste."[14] This liquid position also provided a convenient slippage between moral and industrial concerns. The MPPDA hoped that the government's approval of its business practices would translate into the freedom to regulate its own standards of censorship under the NRA.

Incorporated under Article VII, General Trade Policy Provisions, of the NRA Code, the morality clause read as follows:

> *Part 1.* The industry pledges its combined strength to maintain the right moral standards in the production of motion pictures as a form of entertainment. To that end the industry pledges itself to and shall adhere to the regulations promulgated by and within the industry to assure the attainment of such purpose.

> *Part 2.* The industry pledges its combined strength to maintain the best standards of advertising and publicity procedure. To that end the industry pledges itself to and shall adhere to the regulations promulgated by and within the industry to assure the attainment of such purpose.[15]

A number of groups protested the clause. The Independent Producers, Distributors and Exhibitors Code Protective Committee, for example, argued that "the nebulous character of the pledge set forth . . . is, in itself, evidence of an utter lack of sincerity, and is assuredly and completely ineffectual."[16] The MPPDA, however, took advantage of opposition to the morality clause by construing this stance as an opposition to motion picture morality in general. Thus the Hays Group was able to twist opposition into a reassertion of its own morally superior position.

During the NRA Code hearings in Washington, D.C., in September 1933, Will Hays opened the motion picture industry's presentation of briefs by insisting:

> No section of this code is more important than the morality section. . . . When the whole industry becomes thus pledged, I am confident that we will further progress towards the solution of one of our most difficult problems.[17]

Hays did not specify what the "difficult problem" was. While he undoubtedly was referring to the problem of maintaining a certain level of film quality/morality, the statement carries several subtexts. First of all, the Hays organization was plagued by industry groups that kept raising the issue of unfair trade or labor practices. Independent exhibitors, for example, were against the block booking provision of the producers' proposed NRA Code because this practice forced them to use films and advertising materials that were unsuited to their usual clientele. An additional problem was posed by citizen reform groups. Reformers who had been complaining to the Hays Office for several years about the low level of motion picture morality now joined the exhibitors in their fight against block booking. Double bills, they argued, interfered with local censorship activities and made it difficult for groups to exert pressure on local exhibitors; though "one feature might be approved, the other might be of a type deemed by these groups to be unsuitable for juvenile or family trade."[18] Reformers also objected to the salacious advertising displayed in conjunction with undesirable films, and requested that "more teeth" be put in the Hays advertising code.[19]

In an effort to appease reform groups (and to ease the pressure being exerted by NRA administrators on this matter), the MPPDA made minor adjustments in the block booking provision of the Code.[20] But it did not alter the morality clause. Neither did the Production Code Administration respond to reformers' requests for the enforcement of strict moral regulations. Though the Hays group cultivated public support throughout the latter months of 1933 by maintaining its "open door" policy toward reform groups, it was not until the Legion of Decency was formed in April of 1934 that Hays issued a crackdown and actively enforced the guidelines of the Production Code. This delay indicates that the MPPDA was committed to the public's welfare (or demand for moral pictures) only when forced to do so. In the meantime, the MPPDA used the issue of morality during the NRA Code debates as a decoy to further its own interests. The ultimate achievement of the Hays Office may have been just this: the ability to shift attention away from unfair trade and labor practices while emphasizing, but doing very little about, film morality. The primary evidence of this success is the Production Code's usurpation of the NRA Code in historical discourse. Though the term "the Code" at one time referred to both the NRA Code and the industry's morality code, it eventually came to signify only the latter. Hays's words were thus prophetic. The morality clause, inasmuch as it represented the "morality" of industry self-regulation, did become the most important part of the NRA Code.

Actors' (Im)morality

Will Hays's assertion that Hollywood provided the best wages and working conditions in the world was challenged by nearly every industrial labor group during the NRA period. Thousands of discontented workers staged or threatened strikes, and studio executives were confronted with numerous demands for union recognition under the collective bargaining provisions of the NRA Code. At the Code hearings in Washington, sixty different labor groups appeared before NRA administrators to state their cases for minimum wages and maximum hours. After all the testimony was in, industry leaders were unable to convince government officials of their fair-mindedness in this area of controversy. Though the Roosevelt administration had sided with major producers on issues of trade, its decisions regarding employment practices sided with labor. The open-shop policy was condemned, better working conditions were outlined in specific detail, and, on the eve of the Code's ratification, special allowances were awarded to the industry's creative workforce.

Industry leaders may have lost the NRA battle against labor, but they did not lose their position of dominance over it. Given the labor power differences between the two groups, management was able to maintain the upper hand in negotiations, and its regular refusal to cooperate with labor met with only minimal retribution from government officials. More significantly, industry leaders held a discursive power that was unavailable to labor. While the voice of labor was generally restricted to trade journals and union publications, studio managers faced no such restrictions. Their access to the national media—made possible through vast financial resources; shared capitalist ideologies; or direct tie-ins with newspapers, radio stations, or fan magazines—permitted them to reach and persuade a coast-to-coast audience.

Top-level public relations efforts of this era generally focused on the producers' commitment to the quality and morality of motion pictures, and ignored, whenever possible, any mention of labor. When issues of labor received public attention and became impossible to ignore, the studios rewrote labor discourse in ways that accentuated the producers' moral leadership or patriotic commitment. Actors were central to this revisionist discourse. Since actors were the most visible group of workers in Hollywood, and were the live commodities that embodied the discourse of entertainment, industry leaders had a considerable investment in textual and extratextual representations of actors that fit with studio publicity goals. This meant that, ordinarily, actors were represented discursively as talented performers, charismatic personalities, idols of consumption, civic-minded individuals—as anything *but* laborers. Thus, when actors asserted themselves as laborers during the NRA period, industry leaders were forced to rewrite their actions in ways that maintained the structure of actor–producer relations while reaffirming a discourse of entertainment.

Once again, the impetus for this public relations effort came from the MPPDA. As an organization that was formed, in part, to deal with the problem of stars, the MPPDA understood the sensitivity and urgency of this matter. Producers also knew that the strategies they had employed in 1922 would not be effective in solving the problems they were encountering in 1933. When Hays was brought in to cleanse the image of Hollywood in the wake of the Fatty Arbuckle affair and other scandals in 1921, he enacted a crusade against sensationalism. Hays's first plan of action was to convince the studios that "stars must behave themselves" and that "cheap publicity about stars was dangerous to the industry."[21] Thus, more restrictions were placed on actors' off-screen activities, and studios were careful to select press representatives who would promulgate the new obsession with respectability. In an effort to squeeze out the scandal hunters who turned innocent stars into "victims of the press," the Hays group also persuaded press associations and syndicates to open offices in Hollywood and work cooperatively with studio public relations people.[22] As Alexander Walker summarizes, the producers' "dependence on newspapers and magazines to project favorable images of their stars

[was] only exceeded by [their] anxiety to protect the latter from the unfavourable attentions of writers and reporters." These trends not only indicated the producers' mistrust of the press (and fear that their stories would lead to external censorship), but their "assumption of the right to control an artist's whole life."[23]

In 1933, when actors asserted the right to control themselves and their work, studio executives were forced to adjust their strategies. But, as a congressional investigative report of the 1922 Hollywood scandals suggests, the adjustment was only a matter of emphasis:

> At Hollywood is a colony of people where debauchery, riotous living, drunkenness, ribaldry, dissipation, free love seem to be conspicuous. . . . Some of them are now paid, it is said, salaries of something like $5,000 a month or more, and they do not know what to do with their wealth, extracted from poor people in large part by 25 cents or 50 cents admission fees, except to spend it on riotous living, dissipation, and "high rolling."[24]

In keeping with their pledge to motion picture morality, the studios continued to deny any charges of "riotous living" associated with Hollywood and its stars. But producers found the charge of exorbitant salaries beneficial in constructing their public version of labor discourse. By restricting actors' complaints to the single issue of salaries, producers were able to ignore the more pressing issues involved in actor–producer conflicts. This strategy was consistent with the prevailing explanation of the Depression and its onset; that is, the Depression was "a passing malfunction in an otherwise efficient system, brought on . . . by greed."[25] In response to actors' claims that they were the victims of studio labor policies, producers thus claimed that they themselves were victims of an extravagance gone awry. Producers argued not only that the spiraling costs of stardom had caused the industry's financial problems, but that the stars' refusal to reduce their salaries indicated a greed and impropriety that was out of step with the national recovery program. Thus, after years of protecting actors from charges of immorality, producers now found it profitable to accuse them of immoral behavior.

The extent of the MPPDA's direct involvement in coordinating this strategy is difficult to ascertain. The collected papers of Will Hays, for instance, yield no concrete evidence of his (or his organization's) involvement. But in light of the otherwise copious and detailed nature of the collected papers—they include everything from dinner invitations to annual president's reports to the MPPDA—the absence of memos or reports concerning actors' labor activities is conspicuous. The Hays Office was clearly interested in the effect that actors' labor organizing might have on public sentiment. Evidence of this interest can be found in the numerous newspaper clippings included in the Hays papers. These clippings (primarily from local newspapers across the country) had been pasted together onto sheets and submitted as research reports to the MPPDA president. Most of the news reports concerned the actors' unionizing attempts during October of 1933. Also during this month, a five-page report on newspaper dailies addressed to W. H. Hays from K. L. Russell summarizes newspaper stories, editorials, and letters on the topic of actors' salaries. Russell concludes in his preface to the daily report (entitled "Movie Salaries and Press Agents") that the issue of salary limitations "has come back to plague the industry" and that many Americans question "the right of the government to limit [actors'] financial returns."[26]

A final piece of evidence indicating the MPPDA's interest in counteracting actors' labor activities is an anonymous memo (dated October 13, 1933) that summarizes a press conference held by NRA administrator General Hugh S. Johnson on the topic of salaries. What is particularly significant about this memo is that the bottom portion of it has been carefully torn off. The first few words of this removed section are still (though barely) visible: "You might incidentally tell . . ." The fact that this portion of the memo has been removed (and it appears to be the only example of "damaged" or tampered material in an otherwise pristine collection of papers) suggests that Hays or the MPPDA wished to hide any involvement in this matter. Whereas the newspaper clippings and daily reports indicate a (safe) concern with and awareness of actors' activities, the inclusion of policy decisions or directives against those activities would have presented the MPPDA in an unfavorable light.

The accusation that Hays may have censored his own papers is perhaps unfair; there may have been no incriminating evidence to censor. Since MPPDA interests were equally represented by the Academy of Motion Picture Arts and Sciences, the Academy could, and did, serve as a convenient mouthpiece for the MPPDA and often permitted the Hays Office to avoid formal involvement in labor issues. The individual studios could also rely on their publicity departments to voice the concerns of the MPPDA without implicating the organization or defaming its public image. In addition, the more than two hundred news correspondents accredited by the Hays Office provided a valuable service to producers by printing stories about star salaries that reinforced their discourse of morality and austerity politics.[27] Thus, while distancing itself, the MPPDA was able to wage (or reap the benefits of) an effective discursive campaign that placed actors' demands and activities into question.

Although this campaign did not ultimately convince government officials to legislate stars' wages, gaining public sympathy on the matter was not difficult. As studio executives predicted, "Some 12,000,000 jobless people [did not] enjoy reading about fabulous picture incomes."[28] Or, at least they did not enjoy reading about such wealth without an accompanying critique. Such speculation about the public's reaction to star salaries found its way into fan magazines, one of the most important venues for reinforcing the studios' discourse on stardom. For example, a 1933 article in *Screen Book* entitled "Figuring the Stars' Salaries" notes that for the first time in Hollywood's history, "movie-goers *en masse* are demanding an explanation for bigger stellar salaries."[29] The article opens with a fan letter allegedly received by a prominent, though unnamed, actor. The fan states that she and countless others are troubled by the stars' phenomenal wages in an era when picture houses are closing and so many people have fallen upon hard times:

> I don't begrudge you your fine salary, but don't you think all *big* salaries might be lowered—maybe to $1,000 a week? If what the paper says is true, you earn in one week *five times what most of us earn in a year!*[30]

Asked to respond to the fan's concerns, the actor initially attempts to defend the high salaries of stars. He explains that the professional career of most stars is condensed into only a few years and that, even at one's height, a large percentage of a star's earnings goes to agents, business managers, and upkeep of the extravagant lifestyle expected of stars (by both the fans and the studios). By the time stars "shell out to charity" and pay taxes to Uncle Sam, they are left with only a modest income. This defense takes on a curious twist, however, as the actor states that he was better off in the days when he made only $750 per week and only an agent was "bleeding" him for money. Adding that "real artists and troupers don't require riches for happiness," he concludes that "a general salary reduction to $1,000 per week (in case the government sets such a top on movie salaries) might help rather than hinder stars' screen careers."[31] The actor also argues that handling an exorbitant yearly income is "too much responsibility for anyone with an actor's temperament," and thus may be damaging to one's screen work. As evidence, he states that Clark Gable, Jimmy Cagney, and Ruth Chatterton were all "more arresting screen personalit[ies]" when they were making less than $1,500 per week.[32]

The *Screen Book* article tells us that other "wise and enduring favorites" such as Douglas Fairbanks, Mary Pickford, and Richard Barthelmess, have also found that "big salaries aren't absolute essentials for stellar happiness and artistic progress."[33] Meanwhile, surrounding the text of the article are pictures of the ten top box office draws with their names and weekly salaries printed underneath. Because these stars are not mentioned elsewhere in the article, their publicity stills take on the appearance of "mug shots." They expose and indict the guilty parties who have not realized the detrimental effect of high salaries on the box office and on their own careers. This is a significantly different representation of salaries than that found in a somewhat earlier *Screen Book* article entitled "How Stars Spend Their Fortunes," which simply fetishizes the glamorous lifestyles of stars.[34] In contrast, the NRA Code–era article constructs a moralizing discourse. The question posed in the article's subtitle—"Will the Government Cut Hollywood Salaries?"— prompts the fanzine to provide evidence of salary excessiveness while chastising those stars who might oppose the government's (and the studios') attempt to include a salary-fixing clause in the NRA Code.

For those actors who were not as conciliatory as the unnamed (fictitious?) actor in the *Screen Book* article, fan magazines took a more direct approach in reminding stars of their obligations to the studios. An editor of *Screenland*, for example, felt compelled to send a warning to Ann Dvorak and to "all sulky Hollywood girls" who bucked the moral system of checks and balances that governed the star system.[35] Actors are admonished for complaining about salaries and referring to producers as "slave drivers" during their interviews with the press. Dvorak, specifically, is reminded that although she was a huge success in *Scarface* (1932), she has not yet "arrived." She has not attained the stature of others (such as Joan Crawford, Greta Garbo, and Helen Hayes) who put in years of hard work to build a solid career and earn the public's respect. They were "wise enough or humble enough" not to rebel, says the editor:

> You, Ann Dvorak, are not yet important enough to get away with it. And when you are important enough, you won't want to. The motion picture industry is bigger than you are. It can get along without you, but you can't, excuse me, get along without it. Because no other profession in the world can give you so much.[36]

The editor tells Dvorak that the $250-per-week salary she makes is not much by Hollywood's standards, but it is more than she ever made before. If she wants career longevity, Dvorak must realize that "it's not necessary to make big money fast."[37] The obvious moral of the story is that stars do not *deserve* the huge salaries they receive from their generous benefactors.

This article could not have duplicated the producers' ideological stance toward stardom with greater accuracy. It reveals, for example, how an actor's criticism of working conditions or salaries was seen to stem from selfish ambitions and an unwillingness to work as a team member. Oppositions to studio labor practices, in other words, were never treated as legitimate worker complaints, but as outbursts by "balky and recalcitrant players who through temperament or carelessness h[e]ld up production and add[ed] to the cost."[38] To keep the stars in line—and to make an example of them to other stars— studio executives thus favored a system of penalties that forced individual actors to pay (sometimes literally) for their actions, and that kept intact the structure of labor–management relations.

Variety, a trade publication that regularly reported news on labor–management conflict, meanwhile treated player "outbursts" as a comical staple of industry life. The "doghouse treatment," according to one report, is an endless cycle. "Every week sees some picture name standing with his or her face to the wall receiving the school kid treatment." While those who go into the doghouse "usually feel that they are banished from the land of plenty forever," the studios always "relent."[39] A "Behind the Scenes" scoop in *Screen Book* confirms this phenomenon:

> Although Carole Lombard's pay had been ordered stopped by Paramount when she went on a one-woman strike, she was soon returned to good standing. She contended that the role she was loaned to play . . . for WB was unsuited to her artistry and refused to play it. Paramount forgave the charming rebel and turned to other parts for her.[40]

The doghouse cycle is thus treated as an irritating, but harmless game in which actors who cannot control their temperamental natures force benevolent studios to either discipline them or make special arrangements to accommodate them.

Generally speaking, Hollywood fan magazines perpetuated a discourse of stardom that trivialized actors' labor. In addition to *Screen Book*'s "Behind the Scenes" department, *Silver Screen*'s semiregular feature "Watching the Stars at Work" reduced movie production news to a series of anecdotes about the readers' favorite stars. "Much ado about a cockroach and a New

York actor on the Joan Crawford set at MGM!"[41] is a typical example of its reportage. This discourse of labor was meanwhile enhanced by photomontages (e.g., "Movie Eavesdropping" and "Gallery of Romance") that completely divorced actors from their work and accentuated a discourse of individuality. Thus, unlike some of the NRA-era articles that actively rewrote issues of actors' labor into the dominant discourse of entertainment, these forays into the lives and loves of the Hollywood famous simply shifted attention away from labor altogether. In March of 1933, the month of the studio salary waivers, one of the feature stories in *Silver Screen* was "The Fascinating Mannerisms of the Stars." Although the March issue hit the newsstands before the editors could have anticipated the salary waivers, the point I am making is that the fanzine's general policy was to provide readers with information about the stars that avoided political conflict. Thus, as actors organized their ranks into the Screen Actors Guild in the months that followed, readers would be more likely to know about Eddie Cantor's "big bright black eyes" and Ann Harding's "ultra-feminine, pattering way of running."[42]

Forging a twist on the "star image–real person" dichotomy, fan magazines also created melodramatic stories about how one's career and one's life intersected. The outcome was often tragic, since "Hollywood, the place they call *Heartbreak Town*, has brought nothing but sorrow, disillusionment and discontent to thousands."[43] However, a *Silver Screen* exposé entitled "The Price They Pay for Fame" blamed actors themselves, explaining that tragedies occur because actors sacrifice their health, their friends, and their families on "the altar of terrible ambition."[44] While the article depicts the race for stardom as a "cruel, heartbreaking game" in which fame is a "consolation prize," the studio star system itself is represented as an inevitable part of motion picture life; those who are driven by vanity or greed to enter the system can therefore expect misfortune. Other stories contradict this discourse of self-victimization by reporting that Hollywood has changed some actors' lives for the better. "Out of Tragedy to Happiness," for example, charts the story of how actress Helen Twelvetrees surmounted incredible odds to find romance and success in her motion picture career.[45] "I've Learned Tolerance" explains how the tragedy of ill health "brought a poignant quality to Ricardo Cortez's acting."[46] And *Screenland* boasts the headline "Gable! The Movies Saved Him!" As Gable explains:

Instead of wrecking my life, Hollywood has literally saved me; has skyrocketed me into a glamorous, exciting atmosphere. Remember I once worked in factories and oilfields.[47]

The implication of this narrative is that by becoming an actor, Gable left his identity as worker behind him. Of course, what can never be determined is whether Gable himself actually uttered this statement.

Actors maintained an uneasy relationship with fan magazines. Though the fanzines were essential to an actor's rise to stardom, they often printed erroneous or unauthorized information. Actors could afford to respond with amusement as long as the "ballyhoo" was harmless. For example, in a *Screen Player* article by Ann Harding (third vice president of the Screen Actors Guild), she thanks the Hollywood publicity industry for endowing her with talents in fencing, interior decorating, chemistry, and aviation. But where, she asks, can she find the deed to the beautiful estate she supposedly owns, and proof that she owns the three horses that supposedly are kept there?[48] The editor of the *Hollywood Citizen-News* defended the tactics of Hollywood correspondents, arguing that

> a resourceful reporter, finding it impossible to extract honey from a pearl or gold from a block of marble, learns to fabricate something that is acceptable to his readers. What he fabricates is not always pleasing to the subject interviewed. And the extent of his fabrication varies in direct ratio to the policy of his paper.[49]

At times, however, these tactics could lead to scandalous information that was potentially damaging to the actor.

Concerned about the fanzines' policy of fabrication, the Screen Actors Guild "urged a clean-up of certain fan magazines and other publications which carry sensational, stupid, and often scurrilous articles about motion picture players." The Guild's first protest was to the editor of *Liberty* regarding its "Trade-Views" column, which published "misinformation regarding salaries, and flippant gossip."[50] The Guild was also concerned about those editors who would not accept an interview unless it dealt with the sex life of a star, or a photograph unless it revealed "at least 90 per cent of the subject's epidermis." This policy, they argued, "drives writers, who depend for their livelihood on these magazines, to scandal-mongering and vicious attacks on motion picture players."[51] The Guild admitted, however, that it would be extremely unfair to lay all the blame on fan magazines:

> Publicity men, in their efforts to get space for their clients, have given too little consideration to the material printed in that space;

studios have permitted still photographs to be made which have no possible connection with the pictures they are intended to advertise; and finally, players have been careless and often stupid in their conversations with interviewers.[52]

In August of 1934 the Guild endorsed a series of resolutions that had been adopted by members of the Studio Publicity Executives' Committee and the Association of Motion Picture Producers. These resolutions, though never formally put into effect, did endorse a spirit of cooperation among actors, fanzines, and the studios. According to Ann Harding, "Fan magazines and actors draw their incomes from the same source. Both would profit by cooperating on a constructive policy for the benefit of the industry as a whole."[53]

The most beneficial policy for the major studios, however, was the active dissemination of an entertainment discourse that emphasized temperament, individuality, and a privatized notion of stardom. Under normal conditions, and with the help of their Hollywood correspondents, the studios were able to maintain this discursive and ideological upper hand—leaving actors and fanzine editors to duke it out on the sidelines. But when actors began to rebel openly and collectively during the period of NRA Code development, producers were forced to modify their discursive strategies. They realized that while the temperamental flare-up of a single employee could be handled easily through the trade's gossip mill, the rebellion of organized labor threatened to expose the serious nature of the industry's internal conflicts. Forced to deal with actors as a group, the strategy of focusing on individual personalities had to be incorporated into a broader industrial discourse that could maintain control over actors' labor and the star system hierarchy.[54]

One of the most interesting examples of this strategy was a report delivered to the Academy board of governors and branch chairs by Academy executive secretary Lester Cowan in October 1933. Referring to the actors' active and well-orchestrated campaign against the producers' version of the NRA Code, Cowan accused the creative talent of trying to make the industry into a "cheap racket." He urged actors "to stop parading their problems in public, to stop rehashing the past and to forget petty differences, and get down to the business of making good pictures."[55] These remarks placed actors in opposition to the goals of quality production, defining them as uncooperative, temperamental, and—like the independent film exhibitors who opposed the NRA morality clause—*immoral* members of the Hollywood community who were motivated only by self-interest. The actors' rebellion against the NRA salary clause was thus recast as a "petty" issue that was detrimental to industrial unity.

Cowan's report also employed the discourse of unity to admonish actors for splitting their own ranks:

> High salaried people should think a little more of the other fellow who "isn't" working. High salaried talent who threaten a rebellion to get another $1000 salary a week should think of those men and women who will be thrown out of work by that rebellion.[56]

This statement contains four implications: (1) industry leaders were more concerned about screen extras—and could do more for them—than could their fellow actors; (2) a separate unity among actors of various ranks was not possible (since the desired goals of all actors could be achieved only when unified with management's goals); (3) the result of rebellion was always unfavorable to labor; and (4) labor, not management, was therefore to blame for any suffering that occurred. This admonition, however, is based on a (deliberate?) misreading of the SAG's political platform. As SAG president Eddie Cantor proclaimed in his famous speech earlier that month, the Guild was designed to protect the "little fellow,"[57] and this goal could only be accomplished by the "union" of both high-ranking and low-ranking actors. Disturbed by the number of high-priced stars who were being courted by or had already joined the Guild, the Academy's Cowan was obviously directing this message toward them to discourage their involvement in union affairs. At the same time, the fact that Cowan never mentions the SAG by name reveals the extent to which producers refused to recognize an actors' organization outside the Academy.

Going a step further in his attempt to inhibit union activities, Cowan reminded actors of their responsibilities to their country and their fans:

> Those in the industry who shout "strike" should remember that a strike is the weapon of the oppressed, to be used only as a last resort. [The Roosevelt] Administration and the American people are out of sympathy with strikes, and one flinches at their reaction to a strike on the part of persons drawing high salaries.[58]

According to the MPPDA's belief that its wages and working conditions were "the best in the world," oppressed groups did not exist in the industry; thus, in keeping with this moral logic, actors were using the discourse of oppression without just cause. As a way of reinforcing actors' proper role, this statement links their institutional duties to their civic duties through the weapons of patriotism and audience devotion. It meanwhile ignores management's responsibilities to labor and the Academy's acquiescence to management's requests. While the Roosevelt administration was indeed hoping to prevent strikes, it also called upon captains of industry to negotiate with labor and to provide fair contracts for its workers. The Academy, as chief negotiating arm of the film industry, was failing to fulfill this mandate.

Interestingly, Cowan's warning against the reactions of fans carried an implicit definition of those fans: "American people [who were] out of sympathy with strikes." While this definition may have included a large segment of the civil population, certainly a vast number of fans were themselves striking workers or union members. (During the year 1933, an estimated 1,170,000 U.S. workers—6.3 percent of the total workforce—were involved in strikes that affected industries as diverse as textiles and meat packing.)[59] A particular discourse of fandom, however, was crucial to the overall discourse of entertainment. Just as the studios and fan magazines defined and disseminated a discourse of actors' labor that fit with industrial goals, they defined and positioned the fan as someone who was unsympathetic to or uninterested in striking Hollywood workers.

As a means to further counterbalance the publicity that actors' labor issues were getting during the NRA period, the studios turned a spotlight on those actors who enhanced the image of industrial unity. The studios publicized the names of actors who pledged a small percentage of their salary toward the Motion Picture Relief Fund, a charitable organization that had been established by the industry during the Depression to aid unemployed studio workers.[60] Stars' acts of charity outside the Hollywood community also were featured regularly in the media. A radio broadcast of "Hollywood on the Air," for example, included "short dramatic sketches narrating charitable and other commendable activities of the film stars" (such as Marion Davies's donations to children's clinics).[61] Often these reports were cast in terms of an actor's commitment to national and industrial recovery. As *Variety* reported:

> Donald Crisp's entire time these days is devoted to labors of love. During the day he is doing his duty as a citizen, by serving as a member of the Federal Grand Jury, and at night he works for art's sake, as a directorial member of the Academy emergency committee.[62]

The studios' strategy of publicizing the good deeds of their actors was not a new one, but within the context of the National Recovery Administration it became part of an overdetermined effort to construct a discourse of entertainment that coincided with the Roosevelt administration's ideology of national unity. The result was a conflationary discourse whereby an actor's degree of cooperation with the studios got translated into his or her degree of patriotism and support for Roosevelt's policies. This created a double bind for actors who were soliciting Roosevelt's support for collective bargaining while also fighting against the studios' version of the NRA Code. For, within the industry's public discourse of unity and cooperative participation, such split affinities were seen as treasonous and selfishly immoral. The studios' appropriation of the NRA slogan, "We Do Our Part," consequently provided a convenient and powerful discursive tool for managing labor and encouraging obedience to studio policy. "We Do Our Part" meant that one did one's part on behalf of management. Thus, only certain "parts" for actors (i.e., charitable work versus organizing unemployed actors into unions) were acceptable from management's point of view. As I will show in the following chapter, however, the most important "part" for actors was devotion to the ongoing business of producing quality entertainment. The actors' cooperative involvement in film production not only led to industry profit and favorable public relations; in some cases, the film narratives themselves provided industry leaders with an outlet for their revisionist discourse of actors' labor.

5 / Labor and Film Narrative

Film studies seems incomplete without the study of films. Since much of the recent critical work within film studies has been preoccupied with the notion of "text," and with obtaining a semiotic, psychoanalytic, or poststructuralist understanding of textuality, the Hollywood film has been a privileged object of study. Within cultural studies, however, this form of textual analysis has been placed into question. As Richard Johnson notes, textual analysis remains an important current within cultural studies, but the text is only a *means* to an end. It is "no longer studied for its own sake . . . but rather for the subjective or cultural forms which it realises and makes available."[1]

In the context of studying actors' labor and subjectivity, the film text has not assumed primacy for perhaps obvious reasons. That is, in my attempt to break star studies' fetishistic attachment to the actor as object, a certain distance from the film text has been necessary. But even here, the film text takes on an important role. As Tom Gunning argues:

> There are important tasks of film history, such as the establishment of studio policies and the economics of the industry that need not refer to individual films in any textual specificity. But analysis of the individual film provides a sort of laboratory for testing the relation

between history and theory. It is at the level of the specific film that theory and history converge.[2]

Thus, by juxtaposing a single film against the "rationality of a system," historical analysis can reveal the "complex transaction that takes place between text and context."[3]

The importance of filmic analysis becomes clearer in the context of labor and subjectivity when film is viewed not as a text, but as a commodity. For as Marx argued so insistently, "All commodities are only definite quantities of congealed labour-time."[4] The social relations of capitalist production, he says, are embodied in the commodity and can be traced, in the final analysis, back to the commodity-form. According to Martyn J. Lee, "The real significance of the commodity, then, rests upon the fact that it tends to reflect the whole social organization of capitalism at any historical and geographical point in its development."[5] Having explored in previous chapters the subjective and discursive dimensions of actors' labor within the productive domain of industrial relations, I will now, in the final analysis, subject the commodity-form to a theoretical and historical test. For, if Marx is right, these social relations have been embodied in the textual terrain of the film commodity, and labor policies and labor discourses can be revealed through a textual analysis of films.

One problem, however, is that the commodity is never fixed and cannot be coherently defined. Like the notion of text, the notion of commodity must be opened up to include various contextual aspects of commodification. Hollywood, in other words, produces many commodities, including film publicity and, of course, the film star. The film text as commodity therefore accrues a number of "encrustations" (to borrow Tony Bennett and Janet Woollacott's term) that must be analyzed in its conjuncture.[6] Film publicity in particular assumes a critical role in transforming the use value of the film commodity into exchange value through a manipulation of the symbolic aspect of commodity signs. What follows is thus an attempt to analyze the encrustations of commodification, especially the role of film publicity, in relation to several film shorts and two feature-length presentations, *42nd Street* and *Morning Glory*, that were produced during the NRA period in Hollywood.

NRA Publicity

One of the ways that industry could show support for President Roosevelt's national recovery program was to use the NIRA Blue Eagle insignia in its advertising and publicity. This practice benefited industry in return since consumers were encouraged to patronize businesses and buy products (from Macy's department store to Marchand's castille soap) that displayed the Blue Eagle. In addition, use of the Blue Eagle functioned as a ready form of "goodwill advertising." According to Stuart Ewen, advertisers had worked throughout the 1920s to transform the image of the soulless corporation into a bastion of public concern. "While daily life was projected as a flux of disastrous and unpredictable events," corporations were presented with "a nurturing image of permanence which . . . def [ied] the upheavals of day to day existence" and provided security for the consumer.[7] The economic instability of the Depression era, combined with the promises of the Roosevelt administration, thus provided fertile ground for goodwill advertising. Through such advertising, industry posed as the sympathetic friend to the "little fellow," and reassured the public of its commitment to putting America back to work.

The major film studios capitalized upon the goodwill opportunities made possible by the NRA in the hopes that they could reverse the downward trend in movie attendance and film production.[8] The idea of using the Blue Eagle for publicity first occurred in July of 1933, a time when the film industry was busy developing its Code of Fair Competition. According to a report in *Variety*, the studios decided "to hang the NIRA insignia over the box office and to use it generally in advertising and exploitation." Industry leaders felt that by advertising the NIRA and displaying the Blue Eagle they would not only be "the first to tell the public that [they were] for higher wages and fewer hours," but they would also gain "an extra $10 paid in at the box office for every dollar paid out under the Roosevelt movement."[9]

Even before the Blue Eagle insignia was created, however, studio publicity departments had appropriated the language and events of the NRA for their image enhancement. As early as March 1933, Warner Bros. studio employed the "theme" of Roosevelt's bank holiday to announce its spring lineup of films and to assure the public that Warner Bros. would not fall short of its commitment to quality production in the face of a national crisis. "No Shutdown—No Letdown—but A SHOWDOWN in the war against depression!" proclaims the ad (Figure 1). "Whether you're a Democrat or a Republican you'll endorse—WARNER BROS.' 10-WEEK RECONSTRUCTION PROGRAM."[10] The accompanying illustration of a fist being pounded into an open palm suggests that Warner Bros. meant "business." Embedded in this discourse of entertainment, however, is a complete revision of the role that Warner Bros. and

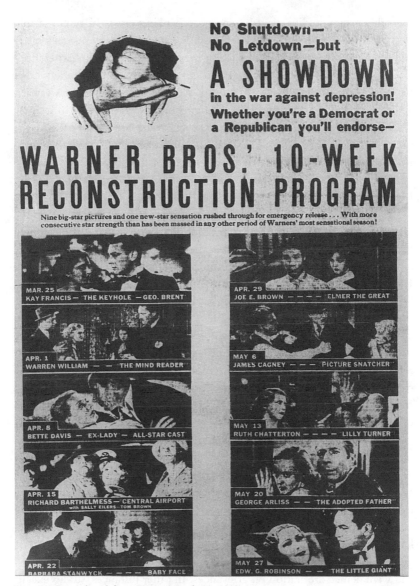

Figure 1. Reproduced from *Variety* (March 21, 1933).

the other major studios played during the banking crisis. In response to a possible shutdown in the film industry, studios imposed an eight-week salary waiver, not a "10-week reconstruction period." And, while Warner Bros. claims "more consecutive star strength than ha[d] been massed in any other period" of the studio's history, the furor over the salary waiver was causing actors to strengthen their own ranks in opposition to the studios. In keeping with its goodwill objective, however, it is important for the ad to leave the impression that Warner Bros., its stars, the public, and the Roosevelt administration are unified by their common goals for national recovery.

Later on, the studios developed "a national campaign along Greater Movie Season lines," which used the NRA as its nucleus for advertising and publicity. One MGM advertisement from this campaign (Figure 2) uses the NRA Code hearings as a theme to promote its patriotism and commitment to public service.[11] The photograph at the top of the ad purportedly represents an assembly of NRA Code administrators and their audience during Code deliberations. Addressing this assembly, Leo the Lion proclaims that "QUALITY cannot be coded!" The implication is that the real business of the film industry—the production of quality entertainment—cannot be legislated by the NRA Code; it can only be created by MGM and its stars. But, like the Warner Bros.' ad, MGM claims "star power" for itself while stars are rendered as passive subjects, graphically and literally embedded within the studio. The Blue Eagle symbol in the lower righthand corner suggests that this "unity" between stars and the MGM name has earned the NIRA stamp of approval.

Some Hollywood employees objected to the studios' use of the NIRA insignia. Members of the International Alliance of Theatrical Stage Employees (IATSE), for example, sent a telegram to President Roosevelt and NIRA administrator General Hugh S. Johnson stating that "the producers [were] using the emblems, or their equivalent, in advertising and publicity . . . at the same time [they were] conducting a campaign to destroy the principle of collective bargaining."[12] They pointed out that, in blatant disregard of NRA guidelines, sound technicians were required to work "unusually long hours" and were denied arbitration on grievances. In other instances, studio executives were criticized for interpreting NRA guidelines in ways that conveniently suited their own interests. The head office at Universal studios, for example, argued at one point that it was unpatriotic to pay employees for meals and overtime: "If our employees remain overtime and put in vouchers for supper money, it will defeat the purpose of the President's agreement."[13]

Figure 2. Reproduced from *Variety* (October 3, 1933).

This memo was criticized for ignoring the studio's role in forcing employees to work overtime and for implying that employees deliberately conspired against the NRA's "shorter hours, more workers" policy. Even those studios that voluntarily fed employees were not exempt from criticism. For whether the NRA was used as an excuse to cut expenses or as a reason to showcase studios' "generosity" to the press, as employee groups contended, the studios were criticized for perceiving the matter of feeding the underemployed as charity and not as a contractual responsibility.

Attempts to prohibit producers from using the Blue Eagle for publicity purposes, however, were unsuccessful. In fact, at the same time that the IATSE submitted its complaint, Harry Warner was seeking approval from General Johnson to produce an NRA screen trailer for theatrical release.[14] To be titled simply *The New Deal*, the film short was designed "to stimulate public interest in the Recovery Program."[15] Within a few weeks, the other major studios had produced their own versions of NRA shorts, and by the end of August 1,000 copies of the eight films were promised exhibition in 8,000 theaters for a total of 64,000 individual showings nationwide.[16] The films (scheduled for an eight-week run beginning September 10) were distributed by the National Screen Service at no charge to the participating theaters.[17] The studios relied on their minor stars or contract players to fill the NRA film roles, but, unlike with other Hollywood films, the stars received little publicity and were not used as box office draws. Subsuming actors within the overall rhetoric of their "We Do Our Part" campaign, studios rather emphasized the content of the films or the fact of the films' existence.

It would be wrong to suggest that, by virtue of their presence in such films, these actors were ideological dupes who merely acquiesced to producers' labor policies. It would be equally wrong to assert that these and other actors did not in some way share the producers' fervent support of the NRA. As American citizens concerned about the nation's economy, and as workers who had much to gain from the NRA's labor policies, actors regularly sponsored and participated in activities that supported the NRA or its principles of community and worker solidarity. The difference between actors' and producers' support of the NRA lay in their differing public visibility, their differing power to mobilize the resources necessary to produce their own positions, and the differing degree of institutional legitimacy accompanying these positions. Both parties, for example, marched in the massive NRA parade in New York City in September 1933. Whereas the Actors' Equity contingent was fairly small and relatively undistinguishable from other groups, the major studios used the opportunity to create a spectacle of entertainment. According to a report in the *New York Times*:

> The Metro-Goldywn-Mayer section created a mild sensation by releasing three large baskets full of pigeons. The Paramount girls, in

blue dancing costumes, were impersonating blue eagles and had to keep their arms outspread, like wings.[18]

The Paramount girls most likely experienced some ambivalence about their role in the parade. While they may have been happy for the opportunity to support the NRA and to boost their career potential, they may also have had some complaints about the working conditions. (Even the *New York Times* noted that they "looked tired.") The studios, meanwhile, represented Hollywood as a discursively unified community, an institutional position that eclipsed any separate concerns held by actors who marched for the NRA.

The studios further capitalized on their productive capacities to establish closer connections with the Roosevelt administration. In the case of the NRA film shorts, the studios volunteered to produce the films for the government's NRA Propaganda Division and to absorb the production costs. Outwardly, this film campaign contradicted the studios' standing policy regarding the use of film for propagandistic purposes. According to the 1932 MPPDA Annual Report:

> The function of motion pictures is to ENTERTAIN. This we must keep before us at all times and we must realize constantly the fatality of ever permitting our concern with social values to lead us into the realm of *propaganda*.[19]

It seems, however, that "propaganda" was another one of those slippery terms employed by the studios according to convenience. In this case, "propaganda" was relegated to those discursive representations that conflicted with national policy. Propaganda was not called propaganda when it was good business and when it garnered the support of government officials who agreed to protect the industry's monopolistic practices. According to John C. Flinn, in charge of the film activities of NRA publicity, the motion picture industry "acquitted itself admirably in the patriotic and important tasks undertaken under [the] supervision [of the NRA Propaganda Division]."[20]

Although the industry's NRA film campaign may have ignored the MPPDA's report regarding propaganda (or interpreted the term in a specifically convenient way), it upheld the notion that the primary function of motion pictures is to entertain. Though the NRA films were part of a propaganda campaign, they were produced not as documentaries, as one might expect, but as classical realist narratives or entertainment spectacles. MGM's contribution, for example, featured Jimmy Durante singing a humorous rendition of "Give Me a Job" in a film short bearing the same title. Other films or "dramatic featurettes" drew from the romantic comedy or domestic comedy genre. One example of this, the Fox film *Mother's Helper*, is described by *Variety* as follows:

> El Brendel tries to explain in a Weber and Fieldian manner how his working only 40 hours weekly will give another man employment. When his wife, Zasu Pitts, wants to know if the NRA affects housewives, Brendel explains he has attended to that and brings in the hot looking Esther Muir, explaining that in the future she'll take care of half of Miss Pitts' wifely duties. Miss Pitts conks Brendel for the fade out.[21]

Within a generic tradition of personalizing social issues, the gender politics of the NRA period are made more palatable (i.e., entertaining) for audience consumption. Within the studios, however, female employees were especially hard hit by the NRA's "shorter hours and more workers" policy. In May of 1933, several thousand female employees, mostly secretaries and stenographers, signed petitions against a bill proposed by the California legislature that would limit the work of all women in the state to an eight-hour day. The major studios also objected to the bill, arguing that the film industry was not like other industries, and that "the emergency nature of much of the studios' vital activities," such as those found in the story department (which employed 1,000–1,500 women), required special consideration.[22] According to a report in *Variety*, however, the studios later praised the passage of the bill because the reduction of work hours for female secretaries and stenographers allowed them more time to concentrate on "domestic bliss" and food preparation for their husbands, thus making everything "hotsy-totsy, thanks to the indigo spreadeagle over the hearth."[23]

Warner Bros.' film short, *The Road Is Open Again*, adopted a more serious tone toward NRA events. According to *Variety*'s description of the film:

> [Dick] Powell is a young composer trying to write music for an NRA song. Visions of Lincoln, Washington and Wilson appear over the piano to advise him. Each tells of his efforts to guide America and admit that President Roosevelt is on the right track.
>
> Powell, through their inspiration, writes the number, "The Road Is Open Again." He sings a verse and its chorus, steps to the front of the curtain and invites the audience to join him.
>
> [The p]icture dissolves into a series of industrial scenes throughout the country with the chorus of the song superimposed on the scenes.[24]

As the most blatantly propagandistic film of the NRA film series, the Warner Bros.' contribution acquired the look and feel of a paid political announcement for the Roosevelt administration. (*Variety* even predicted that the film's "stirring march" would become the theme song of the NRA.) At the same time, the film and its title song served as a form of goodwill advertising that asserted the studio's own economic optimism and authority. Thus, in its sponsorship of the ad/film, Warner Bros. not only displayed its support for the NRA, but celebrated the studio's role in leading the nation to recovery.

Though the NRA films confronted public concerns and offered corporate reassurance in the guise of patriotic entertainment, they were only a temporary and blatant form of propaganda aimed at winning public approval. The more subtle and more profitable means of maintaining public support was achieved through the studios' feature-length films, because these films remained the primary source of contact between fans and the industry and provided the basis of continuity for the production-consumption cycle. While most of these commercial films were of little relevance to the current conditions of crisis, they nonetheless offered a kind of reassurance. Through their sensational plots, elaborate sets, and star attractions, feature films packaged public fantasies and promulgated the discourse of entertainment necessary to studio survival. There were several films released in 1933, however, that directly commented on national affairs or that indirectly voiced the studios' corporate policies. Using the events of the NRA period to reinforce the Hollywood ideology of entertainment, these fictional narratives also provided studios with a powerful outlet for voicing their views of labor, enabling them to shift public attention away from actor–producer conflicts and to rewrite labor issues into a discourse of entertainment.

42nd Street

The best-known feature films of the NRA period that actually incorporated the Depression or the NRA as themes are the Warner Bros. musicals *42nd Street*, *Gold Diggers of 1933*, and *Footlight Parade*. These films have received a great deal of scholarly attention, partly because the first two films of this trilogy were top-grossing films for Warner Bros., and, according to some, helped to reinstate the musical as a respectable and profitable genre.[25] Film scholars have examined these films for their gritty realism, their escapism (attributed primarily to Busby Berkeley's production numbers), or their spirit of cooperation with Roosevelt's New Deal.[26] In his article "Some Warners Musicals and the Spirit of the New Deal," Mark Roth argues that the mu-

Figure 3. The Better Times Special promotional tour for *42nd Street* (Warner Bros., 1933). (Courtesy of the Academy of Motion Picture Arts and Sciences.)

sicals are "essentially political" because their purpose is "to come to terms with the questioning of the American Dream and to reaffirm faith in that ideal."[27] However, as films that claimed to inaugurate a "New Deal in Entertainment," the full implication of their political force can only be understood by examining how their restoration of the American Dream was beneficial to the Hollywood image—and studio labor policies. For these musicals are not simply films about the NRA; they are films about the acting profession.

42nd Street, the first of these musicals released in 1933, was preceded by an extravagant promotional tour (Figure 3). On February 21, a troupe of chorus girls and several of Warner Bros.' top stars (Bebe Daniels, James Cagney, Dick Powell, and Joe E. Brown) boarded a train in Los Angeles. The train, called the Better Times Special, stopped at several key cities (Denver, Kansas City, Chicago) on its trek eastward before arriving in Washington, D.C., for Roosevelt's inauguration on March 4; the group's final destination was 42nd Street, New York.[28] The promotional tour was made possible through the cooperation of General Electric, which saw the tour as a good opportunity to advertise its own products:

> The outside of the train [was] leafed in gold and silver with a constant electric sign burning the legend "Better Times" and announcing the Warner picture and the G.E. equipment contained therein.[29]

As the parent organization of RCA and NBC, GE was in a position to authorize hookups between local radio affiliates and the train's broadcast facilities. At each stop, the Better Times' radio broadcast songs by Dick Powell while it advertised GE products, the Warner Bros. film, and the NRA spirit of optimism. During the day, the stars were transported to GE showrooms to demonstrate appliances, and, in the evenings, they appeared at local theaters for premieres of *42nd Street*.[30] In what *Variety* called "one of the juiciest exploitation tie-ups known [to Hollywood],"[31] actors thus served as living advertisements for the studio's products and discourses of entertainment. But within the fictional narrative of *42nd Street*, the presence of actors took on an added significance. Here, actors provided Warner Bros. with an opportunity to comment upon actors' labor and subject positioning within the studio system.

The plot of *42nd Street* is simple. Set as a backstage musical, the story revolves around "puttin' on a show." When the star of the show breaks her ankle, a determined young novice learns the dance numbers and becomes an overnight success. But against the backdrop of this simple plot, *42nd Street* establishes an authoritative commentary on actors' labor. The mechanism for beginning this commentary occurs after the opening montage sequence of New York's theater district as the camera settles on a close-up of an Actors' Equity contract (Figure 4). It reads:

> Jones and Barry, Theatrical Producers, hereby engage Dorothy Brock to star in their musical production *Pretty Lady* . . .

What is interesting about this shot is that a display of the contract is not necessary as a plot element; that is, viewers do not need to *see* the document in order to pick up information essential to the plot. Yet, given its privileged position and prominence after the film's preface, the Equity contract serves as the source of conflict that sets the narrative into motion. Actors' labor becomes condensed into the Equity contract as an overdetermined and isolated sign that signifies a site of tension between actors and producers. The task of the narrative, then, is to find a resolution to this conflict. By using a "genuine Equity contract" (as called for in the script) the film makes the threat of organized labor more visible, and increases the possibility that viewers will sympathize with the terms of narrative closure.

In addition to representing actors' labor in general, the contract specifically represents the Actors' Equity union, an organization that had caused considerable problems for Hollywood management since its arrival in 1919 (see chapter 2). Though the union was divested of bargaining power in 1927

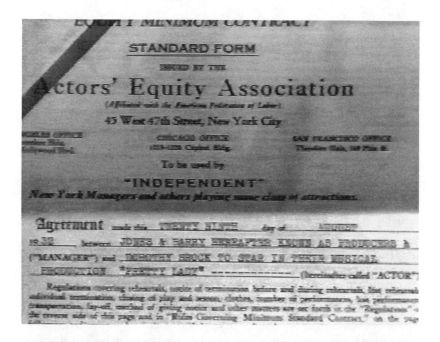

Figure 4. Producer Abner Dillon (Guy Kibbee) inspects an Actors' Equity Association contract (close-up, *above*). (Frame enlargement from *42nd Street*, Warner Bros., 1933.)

when the Academy was formed, Equity still maintained a noticeable presence in Hollywood (screen actors who worked in the theater necessarily retained membership, and many others never bothered to sever their ties with the union), and studios continued to be wary of its influence. In the film *42nd Street*, however, Equity's eradication from Hollywood is finally achieved. Producers return the union to its "proper" origin by associating Equity with the theater. This movement has the advantage of displacing actor–producer conflicts from Hollywood onto the New York stage, and of allowing the film to present an interpretation of actors' unions that is disassociated from inequitable labor practices in Hollywood. Viewers were thus encouraged to identify with an ideology of labor that was distanced from, yet integral to, Hollywood.

Management literally has its hands and eyes on actors' contracts from the beginning of *42nd Street* as *Pretty Lady*'s producer, millionaire Abner Dillon (Guy Kibbee), holds the contract up for inspection (Figure 4). Management's point of view (and, thus, the desired point of view for the audience) is established through a shot–reverse shot technique that focuses first on a close-up of the Equity contract, then cuts to a medium shot of Dillon holding the contract. This perspective is reinforced orally through voice continuity. During the close-up we hear Dillon's voice-over: "Well, of course, I'm not a lawyer. I'm in the kiddie car business." But as the film cuts to the shot of him holding the contract, Dillon is shown speaking: "I don't know much about contracts, but this looks good to me." As Dillon speaks the final clause of his statement, the camera cuts to a close-up of a woman's legs (shown in mirror reflection). It is clear that what "looks good" to the producer is not the contract, but the legs of Dorothy Brock (Bebe Daniels), the show's star. This association discredits the terms of the contract by substituting the star's sexuality for the legal document. As such, the cutaway defines the nature of contracts as payment for a fetish object rather than as procurement of an actor's labor.

A publicity still for *42nd Street* (Figure 5) offers an interesting comparison to the scene that occurred in the film release. In the still photograph, both "characters" are more or less facing the camera, and we do not see the contract exclusively from the producer's point of view. Indeed, the actor, not the producer, is in possession of the contract, and the latter is positioned behind and to the side of the former to see the contract from her visual perspective. While there may be sexual overtones to this photo, the producer is also not directly fetishizing some aspect of the actor's body through the sexual gaze. The different positioning of the characters in relation to the

Figure 5. Dorothy Brock (Bebe Daniels) and Abner Dillon (Guy Kibbee) in *42nd Street* (Warner Bros., 1933). (Courtesy of the Academy of Motion Picture Arts and Sciences.)

contract thus creates a more harmonious representation of labor relations. I doubt, however, that this was the purpose of the publicity photo. It was common practice for publicity stills to be taken on the production sets by studio photographers (who may or may not have known the plot of the film) and not copied from the films themselves. Moreover, since the point of such publicity was to display the starring actors (and not a single, minor actor), it is not surprising that top-billed Bebe Daniels is present in the still. This comparison, however, only underscores the significance of the shot–reverse shot sequence used in the film. Although the harmonious relations in the publicity photo deemphasize the contract (a viewer may not even realize that the object Kibbee and Daniels are looking at is a contract), the foregrounding of the Equity contract in narrative form sets the stage for actor–producer conflict.

Dorothy Brock's response to Dillon ("It's the biggest contract *I've* ever signed—thanks to *you*, Mr. Dillon.") immediately places the actor and producer at cross-purposes since the star defines her relation to the contract primarily in terms of salary. Yet, the coyness of her response—as she peeks over the top of a *New Yorker* magazine, then slips him a knowing (though slightly disgusted) smile—indicates that she has achieved her role and contract through sexual manipulation and thus is implicated in Dillon's understanding of the contract as a sexual agreement. Later on, then, when the star withdraws her sexual attention, Dillon's threat of dismissal is presented as a justified action. In an interesting and effective conflation of discourses, labor–management conflicts are collapsed into gender politics, and the sexual contact between male producer and female star is shown to be more binding than the union contract. In the final analysis, the union contract means nothing more than the producer's definition of it. The film furthermore suggests that actors are better off without one. For when novice Peggy Sawyer (Ruby Keeler) replaces the star, she does not sign an Equity contract. Grateful for the chance to become a star, and willing to do whatever management requests, the success of the nonunion actor far outshines that of the union star.

In *42nd Street*, union contracts obviously carry moral implications. Dorothy Brock, the only actor (i.e., character) in the film to sign a contract, is depicted as petty, gold-digging, and sexually suspect. She is concerned only with salary and status, not with the welfare of others or the success of the show. Peggy Sawyer, by contrast, is sexually pure and uninterested in money. Her only desire is to get her chance on Broadway; she leaves all decisions regarding salary and performance to management, implicitly trusting that she will be well cared for. Sawyer was not the sort of actor who would turn the industry into a "cheap racket" as feared by Academy spokesman Lester Cowan. Nor would this type of actor be likely to sabotage the public relations efforts of the Hays Office (see chapter 4). It is no coincidence, then, that this cooperative and obedient character was represented as the more desirable actor to audiences.

Audience identification with the Sawyer character was further reinforced through the wholesome star image of Ruby Keeler. Far from being a sex goddess, Keeler was perceived as an innocent, working-class girl with high morals, lots of enthusiasm, and only average talent. According to Rocco Fumento,

> There's something so very vulnerable about her, so moronically endearing in her klutziness while she earnestly looks down at her feet as she clumps her way up Forty-second Street . . . But women liked her . . . Old men wanted to protect her (and her incredible innocence) . . . young men wanted to embrace and make her their wife, *never* their mistress . . . and to boys stumbling into puberty she was an angel, a first infatuation . . . a madonna of the musicals.[32]

This image of wholesomeness was exploited in the studio publicity surrounding the Warners' musical trilogy. Advance features sent to theaters, for example, sported headlines such as "Ruby Keeler Just an Old-Fashioned Sweet Girl" and "Ruby Keeler, Millionairess Is as Timid as an Extra." In press kit interviews, Keeler was modest, playing down her theatrical abilities and playing up her homespun qualities.[33] According to a report in *Variety*, Keeler refused to continue production on *Footlight Parade* until the studio provided her with a less revealing costume: "Keeler claimed that she could not do her dance properly, and that she was more interested in giving a satisfactory performance than in supplying s. a. [sex appeal] to picture fans."[34] Though Keeler could have been penalized for her rebellion, Warner Bros. undoubtedly did not want to place the studio's commitment to morality into question, and thus immediately honored her request.

The studio's concern over maintaining a morally acceptable image had also led them to take certain precautions in preparation for the *42nd Street* Better Times publicity tour. They were particularly concerned about how the public might perceive the twelve young women of the chorus troupe who would be traveling across the country in mixed company. According to an interoffice memo,

> The trip in itself is . . . one which may result in unfavorable criticism of the company in the event anything of a serious nature should happen to anyone [*sic*] of these young girls, and it is needless for me [R. J. Obringer] to impress upon you the necessity of strict discipline.[35]

In an attempt to avoid any problems, Warner Bros. included a clause in the chorines' contracts that required them to "promptly and faithfully" comply with the "especially strict rules" set down for the purpose of the tour:

> It will be particularly required that all persons upon such tour . . . shall conduct themselves with special regard to *public convention and morals* and that no action which will tend to degrade Artist or the tour, will bring either Artist or any member of such tour into public hatred, contempt, scorn or ridicule, or which will tend to shock, insult or offend public morals or decency, or prejudice the Producer or the motion picture industry in general, will be tolerated.[36]

Failure to comply with these terms (which remained necessarily vague) was cited as sufficient cause for termination.

Although the chorus girl contracts for the film production of *42nd Street* did not include a morality clause, Warner Bros. still felt compelled to promote the idea that the studio cared about its employees' moral welfare. According to a press kit article credited to the *Brooklyn Daily Eagle*, the chorus girls in *42nd Street* were put on a strict diet, a stricter exercise routine, and an even stricter mode of living. To save their beauteous legs, they were provided with roller skates, on which to glide from their stage to the restaurant. Too, they all live under one roof; must be in bed by 9 o'clock, must be up at 7. Nor is that all. For on the way to and from the studio, the gals are chaperoned by a corps of husky studio policemen.[37]

This connection between morality and strict discipline became an implicit part of every actor's agreement with a studio. Even when morality clauses were not included, every contract stipulated that the artist would perform services "in a conscientious and painstaking manner and in accordance with the reasonable instructions of the producer [and] the reasonable studio rules and regulations of the producer."[38] The notion of what was "reasonable" was, of course, the source of conflict between actors and producers.

According to a Screen Actors Guild retrospective report, actors at this time were faced with unreasonable demands due to unregulated hours and working conditions:

> Actors were required to work almost every Saturday night and often into the early hours of Sunday morning. If a studio closed for a holiday during the week, the actor often would be required to work the following Sunday without pay to make up for the holiday. Meal periods came at the producers' convenience, not necessarily to meet the human needs of the actor.
>
> There was seldom any 12-hour rest period between work calls. Actors often worked well past midnight and then were ordered to report back for work at 7 a.m. Actors were not paid for overtime and no premium was paid for work on Saturdays, Sundays and holidays nor for night work.[39]

During the filming of *42nd Street*, "Daily Production and Progress Reports" indicate that actors (and other employees) were indeed asked to work overtime (including Sundays), and were not always given a twelve-hour rest period between work calls, even though these practices violated the terms of their contracts. During a two-day dress rehearsal for one of the Busby Berkeley numbers, for example, the cast worked seventeen to eighteen hours per day (10:00 a.m. to 2:30 a.m. and 9:00 a.m. to 2:50 a.m.) with only a six-and-a-half-hour rest period in between. Thus, contrary to Warner Bros. publicity, chorus girls were not in bed every night by the wholesome hour of 9:00.

This type of situation is duplicated in the narrative of *42nd Street*. Director Julian Marsh (Warner Baxter) keeps the cast members well into the night during a grueling dress rehearsal for *Pretty Lady*. When the actors become tired and inattentive, he yells: "Not one of you leaves this stage tonight until I get what I want." One elderly actor does leave after fainting, but the others remain until Marsh is satisfied with their performance. Another time, when Peggy Sawyer faints during a rehearsal, Marsh orders the actors to resume their positions. "This is a rehearsal," he says, "not a rest cure!" According to a report by the Chorus Equity division of the Actors' Equity Association, chorus girls at this time worked up to ninety hours per week and were paid below regular wages, if anything at all, for rehearsals. Having little or no money to eat (or no time to eat if the director failed to call a lunch break) would easily have caused a girl like Peggy Sawyer to faint from exhaustion.[40] In the case of *42nd Street*, however, the director was not violating any contractual agreements, because the film implies that only the star of the show has signed a contract. Marsh's actions are further motivated and exonerated by his ill health and poor financial condition; although nearing collapse, he pushes himself and the actors to produce a successful show.

Occasionally, the actors are cynical, or even critical, about their working conditions, as in the following exchange:

—They [the rehearsals] just about kill these youngsters . . . and for what?

—For thirty-five a week—when you can get it!

Indeed, though some theater managers claimed that chorines earned $40 a week and over, statistics compiled by Chorus Equity showed that "the remuneration in most instances was not more than $25 a week."[41] The quoted remarks remained a form of resistance that had little potency, however, in terms of changing working conditions. Because the actors were otherwise portrayed as lazy, insolent, temperamental, untalented, or sufficiently lacking in intelligence, these verbal resistances served as comic one-liners that diffused labor–management tensions and implied that the actors were not to be taken seriously. According to the moral logic of *42nd Street*, the major reason the show became a hit was the director's unrelenting vision and dedication. This idea is reinforced in the final scene through an ironic twist. After the show, as Marsh leans wearily against a fire escape in the alley, he listens to the remarks of people coming out of the theater and hears them praise everyone but himself. ("I can't see that Marsh did a thing." "It's simply having the right cast—that's all!") The implication, however, is that the director deserves all the credit; he was the one who transformed these feckless players into a productive and obedient group of entertainers.

Ultimately, the key to the film's morality is an ideology of labor that combines individual initiative with collective subordination. In his article on Warners' musicals, Mark Roth finds these two notions contradictory, arguing that the film's Horatio Alger version of the American Dream—in which the myth that individual initiative, hard work, and luck lead to ultimate success—is contradicted by the film's production numbers (and its resolution), in which we see many individuals subordinated to the will of a single person—the director.[42] But, contrary to Roth's argument, both individual initiative and collective subordination were essential to reinforcing an ideology of stardom that served management goals. Individual initiative promoted the idea of mobility within the star system hierarchy. When individual initiative was perceived as "hard work," actors low in the hierarchy (like Peggy Sawyer) might be rewarded with an opportunity to move upward; but when individual initiative was perceived by management as a temperamental display or selfish demand, stars (like Dorothy Brock) could be threatened with dismissal or downward mobility. Contained within the myth of individual success, then, is the notion of subordination. Studios merely extended this notion to the sphere of collective work: actors were expected to collectively serve management's (moral) goals instead of forming (immoral) collective bargaining units to protect their own interests.

Although *42nd Street* was conceived, developed, and produced in 1932, before the NRA even existed, Warner Bros. capitalized on the occasion of Roosevelt's inauguration and the popularity of the NRA for the film's release in March 1933. Within this context the film's portrayal of actors served to strengthen studio labor discourse. Through their dedication to the collective production of the show, actors come to represent the collective spirit of the New Deal. This allegorical shift allows the film to ignore or displace the specific problems of actors. In other words, *42nd Street* is no longer a film about actors struggling to survive in their profession, but a film about American citizens struggling toward national recovery. Through this redefinition, the film successfully diverts attention away from actor–producer conflicts while implicitly encouraging viewers to sympathize with management's ideology of labor. As in the NRA film shorts, actors' labor thus becomes subsumed under a collective, industrial discourse that sought unproblematic and profitable solutions to quality entertainment and national recovery.

Though *42nd Street* was the third top moneymaker of the year for Warner Bros., the studio did not feel obliged to share the fruits of collective effort. On the contrary, the studio's profits were accumulated at a time when it sought to curtail actors' wages even more by imposing, and trying to extend, its eight-week salary waiver. Guy Kibbee, who played the lecherous, bumbling producer in *42nd Street*, protested against this situation. In a letter to Warner Bros. studio he argues that, in view of the fact that the producers "refused to comply with the terms and conditions set forth in the Emergency Bulletin" by choosing to extend the salary waiver period beyond the agreed upon eight-week period, he no longer considers himself legally bound to his signature on the salary waiver form and therefore demands that Warner Bros. compensate him for the period of the salary waiver as per his original contract.[43] Kibbee's letter was sent a few days after he received notice from Warner Bros. that the studio had decided to exercise its right to lay him off for a period of three weeks without pay. The studio justified its action by citing the paragraph regarding noncontinuous employment in Kibbee's contract dated May 16, 1931 (the same contract that Kibbee refers to), but its letter does not mention the salary waiver period or Warner Bros.' obligation to honor the original terms of said contract as regards salary.[44]

Some of the *42nd Street* actors posed no problems for Warner Bros. management. As Thomas Schatz has pointed out, Dick Powell was not only well suited to "Warners more economical and genre-based approach to production," but was willing "to work more often and for lower salaries" and to let the studio shape his screen roles.[45] Powell's participation in both the Warners' NRA film short and the *42nd Street* promotional train tour supports the idea that he was amenable to the studio's broader publicity goals as well. His dance partner in the musical trilogy, Ruby Keeler, also fit well into Warner Bros.' publicity scheme. The studio played up the fact that, after a successful career on Broadway, Keeler was getting her "big break" in motion pictures, and thus her story was not unlike that of Peggy Sawyer in *42nd Street*. In one publicity article, for example, Keeler is quoted as saying that landing the lead role in *42nd Street* was a stroke of "luck" and that she felt as timid in her first film as Sawyer did in her first stage production.[46] One difference between Keeler and Sawyer, however, is that the former signed a multiyear contract that guaranteed substantial yearly increases in salary.

While Warner Bros. may have tried to exploit certain connections between the film and its actors, aspects of labor–management conflict within the industry were rewritten or made invisible by the film's narrative discourse about labor. It is ironic, for example, that Bebe Daniels plays the role of a prima donna who has a contract dispute with her producer, because several months prior to the filming of *42nd Street* Daniels was engaged in a lengthy contract dispute with Warner Bros. Daniels refused as many as eight stories from Warners on the basis that they were unsuitable and that the studio had not given her "the proper opportunity to prove her drawing power."[47] She finally took it upon herself to submit a story for consideration, but Warners refused to produce it on her terms (which included choice of director and final approval of the dialogue). For several months Warners and Daniels's agents made compromises and counteroffers. Because both parties wanted to sever the contractual arrangement, the issue was no longer which picture she would play in but how she could finish out the terms of her contract. In an affidavit regarding Bebe Daniels, Darryl F. Zanuck concluded that Warner Bros. "had conscientiously endeavored to compromise with her and she had stalled until [they] were over a barrel."[48] In the end, Daniels managed to get a contract that was, for the most part, on her terms.

George Brent, the actor who portrayed Dorothy Brock's boyfriend in *42nd Street*, became engaged in a lengthy contract dispute with Warner Bros. several months after the film's release, but he was not as fortunate as Bebe Daniels. Brent's refusal to report to the set of a movie for which he was scheduled resulted in his suspension. Warner Bros. also sent a letter disclosing Brent's action to nine different studios and to the Association of Motion Picture Producers.[49] While this letter put Brent into a bind by informing other studios that he was not a free agent and not available to sign elsewhere, it also was meant to protect Warner Bros. against star raiding. Warners was a leader against star raiding, at one time even calling a special meeting of the Hays organization to investigate the alleged raiding practices of 20th Century studio.[50] Meanwhile, screen actors were fighting against the producers' anti-star raiding provision of the NRA Code, a document whose process of conflict and negotiation tells a very different narrative of actor–producer relations than that found in *42nd Street*.

As Jane Feuer notes, "The musical turns its self-reflective technique to its own purposes," oscillating between demystification and remythicization to reproduce "the myth of entertainment."[51] The backstage musical, in particular, purports to foreground the secrets of its making by providing viewers with a behind-the-scenes look at the production process. But this demystifying technique only sets the stage for a revisionist discourse whereby the film mythologizes the economic and stylistic forces of production—including the role that labor plays in this process. Thus, when Warner Bros.

foregrounds the acting profession and its working conditions in its backstage musicals, actors' labor becomes subject to commentary and reformulation. This strategy allows for a double reinforcement of the studio's version of labor discourse by promoting management's point of view while simultaneously discounting or silencing alternative points of view from actors.

I do not wish to imply that the Screen Actors Guild version of labor–management relations describes the "real" situation and that the studios created a "false" version of labor through their publicity and film narratives. Hollywood discourses of entertainment were certainly real enough to actors and affected actors' labor in concrete ways. The point is that in the discursive struggle over the meaning of actors' labor, studio management had a more advantageous position from which to speak. The studio additionally benefited from the films' widespread distribution, which guaranteed that the public would have much greater access to the films than to information that would have challenged the studio's position on labor–management relations. In light of these factors, and given the immense popularity of the films, it appears that audiences accepted, at least on some level, the studio's version of labor discourse.

Morning Glory

The Warners' musicals did not singlehandedly rewrite actors' labor into a discourse of entertainment during the NRA period; they merely stand out because their representations of the acting profession had direct tie-ins with economic and political events. Other films of this period promulgated the familiar discourse about stardom and complemented the studios' public relations efforts on other fronts. One example is *Morning Glory* (1933), a top-grossing RKO picture described by one critic as "*42nd Street* without music, dancing or Ruby Keeler."[52] Katharine Hepburn plays the lead role of Eva Lovelace, a stagestruck young woman from Vermont who goes to Broadway to fulfill her destiny as a great actress. When her "natural talent" is not immediately recognized, the actress perseveres; even when she is penniless and homeless, Lovelace retains an undying (if not naive) faith in the theater and in herself. This dedication and reverence toward acting become the key to her success as she finally is rewarded with a chance in a major Broadway production.

Unlike *42nd Street, Morning Glory* is not so much about collective subordination as it is about individual initiative and artistry. This tribute to individualism is established in the opening scene as Lovelace enters a theater

Figure 6. Louis Easton (Adolphe Menjou) and Rita Vernon (Mary Duncan) in *Morning Glory* (RKO, 1933). (Courtesy of the Academy of Motion Picture Arts and Sciences.)

lobby and stands in awe before the portraits of great stage actors such as Sarah Bernhardt, Ethel Barrymore, and John Drew. Lovelace's own budding greatness is revealed through her subsequent encounters with Rita Vernon (Mary Duncan), an actor who has just landed the lead role in the comedy production *Blue Skies*. Like Dorothy Brock and Peggy Sawyer in *42nd Street*, Vernon and Lovelace represent two types of actors with two different attitudes toward their work. Vernon is the stereotypic gum-snapping, peroxided chorus girl who has made it to the top through her streetwise (i.e., sexual) manipulations (Figure 6). Vernon is a careerist, not an artist; she is concerned primarily with status and money, and acting appears to be a glamorous means to achieve these ends. By contrast, Lovelace has naturally colored hair along with her natural talent (Figure 7); she is an artist whose passion is reserved for the stage. By setting up such a distinct comparison between these two actors, the narrative must resolve the conflict they embody. The turning point comes when Vernon demands a contract.

The only reason Vernon agreed to play the lead in *Blue Skies* is that her producer, Louis Easton (Adolphe Menjou), promised her that she could have her choice of plays thereafter. The play she chooses is *The Golden Bough*, a dramatic piece that she feels will allow her to develop the image of a serious actress. But on opening night, fifteen minutes before curtain time, Vernon confronts Easton with contract demands. We learn that for four years Vernon has never had a contract; instead, actor and producer had a verbal agreement

Figure 7. Joseph Sheridan (Douglas Fairbanks, Jr.) and Eva Lovelace (Katharine Hepburn) in *Morning Glory* (RKO, 1933). (Courtesy of the Academy of Motion Picture Arts and Sciences.)

whereby she would be "reasonable," Easton would "do the right thing," and they would "take each other's word." Vernon reminds Easton that while she has never "talked salary," she has made a fortune for him. In addition, "being reasonable" has meant that Vernon never asked for what she *really* wanted—until now. Vernon makes several demands. She wants her name in electric lights; a run-of-the-play contract for *The Golden Bough* in New York and on the road; $1,500 a week; half the show's profits; and a percentage of the motion picture rights.

Vernon feels confident that Easton will agree to her demands, since she has placed him in a tough spot. But Easton responds with anger, threatening to report Vernon to the Actors' Equity Association and asserting that he will make certain she never works again. However, when Vernon says that she will not appear on opening night, he fears that his entire investment in the show will be lost if he refuses to meet her terms. Just as Easton is prepared to give in, playwright Joseph Sheridan (Douglas Fairbanks, Jr.) urges him to take a chance on the understudy. The understudy, of course, is Eva Lovelace, who, unbeknownst to Easton, had been hired by Sheridan for the part. Easton agrees to give the novice a chance and reports back to Vernon: "Since you've decided to act in this most *unprofessional* manner . . . I've decided to let you do exactly as you please." Vernon's exit simultaneously marks her eradication from the play/film and the producer's victory in the contract dispute.

Lovelace is a success in the play, but the film audience does not witness her performance. In fact, aside from a few brief shots that indicate that Lovelace has taken interim jobs in variety shows, *Morning Glory* never represents the actors at work. Unlike *42nd Street*, which must portray actors at work in order to make the connection to Americans working toward economic recovery, *Morning Glory* abolishes any notion of actors as laborers. This construction allows the film to explore the artistic side of the acting profession while separating labor from art. The film still places actors in a subordinate position, however. As the film suggests, actors who perceive the acting profession as a spiritual or artistic calling do not question labor practices or demand (better) contracts. Artistry is furthermore linked to morality in the eyes of producers. Vernon, a mediocre talent with questionable moral habits (drinking, sexual propositioning), is finally dismissed when she attempts to define herself as a worker. Although the film intimates that Lovelace and Easton have a sexual encounter (he takes advantage of her at a party when she is drunk), she retains her moral sensibilities by defining herself in nonthreatening terms as an artist. The key to moral relations between actors and producers is thus a matter of who is initiating what and who plays the subordinate role.

After her performance on opening night, Lovelace's one desire is to make Easton proud of her—a desire complicated by the fact that she has fallen in love with him. But Easton puts her new role in perspective by explaining that she is now under his *professional* wing:

> —It's going to be difficult, my dear, because I'm a difficult task master and I'm likely to make a pest of myself. But you're too valuable to ever get out of my sight. . . . That's why you're coming to my office tomorrow and signing a contract to play this part as long as the public wants to see you in it.

Easton is no longer threatened by the idea of a contract once he is able to initiate it on his own terms. Although a contract was not "necessary" before, it now establishes a (task) master–slave relationship that allows Easton to protect his investment. (Easton refers to Lovelace as "the most valuable piece of theatrical property I have ever owned.") The contract is not intended to protect the actor's investment of labor, nor is it designed according to NRA guidelines concerned with protecting actors as a group.

When a producer has the power to grant (or refuse) contracts according to his own convenience and terms, actors are divested of collective bargaining power. It is interesting that, like *42nd Street*, the narrative of *Morning Glory* recognizes the existence of Actors' Equity but attributes very little power to it. In this case, Easton ironically assumes that the actors' union will protect *him* from temperamental actors and that he has the power—and the right—to intervene between an actor and her union. But while Easton goes so far as to threaten Vernon with retaliation from the Equity union, it is unclear on what grounds Easton feels the union could logically retaliate. Perhaps because Vernon had cooperated with the producer and forfeited an Equity contract for four years, she would somehow be ineligible for union membership in good standing. More likely, if Vernon was already an Equity member and was working without a contract, she could automatically be suspended from the union for undermining its objectives. Equity regulations stated that "an Equity member may play only in companies where every member is a member of Equity in good standing."[53] The union felt that stardom, especially, involved certain responsibilities as well as privileges. If stars had "any hand at any time in breaking down the conditions of the standard contracts they [would] be haled before the Equity Council."[54]

As the Equity policies suggest, a union-minded actor would be concerned with the treatment and fate of other actors in a show. Lovelace, however, never asks what happens to Rita Vernon. Easton has successfully divided the two actors from each other and prevented any contact that might cause them (and other performers in the show) to work together against his interests. Lovelace also does not question her own fate. After her successful debut, a seasoned actor warns her that she could become a "morning glory" ("a flower that fades before the sun is very high"), but Lovelace responds that she is not afraid of fleeting stardom. Lovelace thus accepts an ideological discourse of stardom that insists that even the most gifted artists must subordinate their talent to a producer-dominated structure of labor with no guarantee of job security or control over their image. According to a fan magazine's assessment of the situation for screen actors: "Idolatry today, Isolation Tomorrow. That's the Dangerous Paths [*sic*] of Hollywood Glory."[55]

Morning Glory represents the star system as a heartless but immutable structure in which individual artistry breeds competition among actors, and success is often determined by an actor's degree of cooperation with producers. Actors experience problems when they are unable to cope with the system, when they expect too much from it (e.g., Rita Vernon), or when events in their personal lives conflict with their professional goals. For example, when Lovelace finally falls in love with Sheridan, the playwright, she realizes that she must learn to juggle a career and a family life. The film's depiction of the difficulties encountered in the acting profession are not enough, however, to undermine the romanticization of stardom. Though Lovelace occasionally goes without food or income in her struggle to reach the top, she nonetheless becomes a star. The moral of the story appears to be that if one has talent and is dedicated enough, one will achieve one's destiny. How long one remains a star is immaterial.

Hollywood studios sent this same type of mixed signal to young hopefuls trying to enter the acting profession. Though studios discouraged young girls from the Midwest from traveling to Hollywood in hopes of landing a screen test and being discovered, the rags-to-riches narratives found repeatedly in fan magazines and films such as *42nd Street* and *Morning Glory* reinforced a discourse of stardom that seemed to place success within the reach of starry-eyed novices. A typical example is the article "Extra Girl Gets Her First Close-up!" in which writers for *The Hollywood Reporter* imagine what Jean Harlow must have been thinking before her big moment arrived:

> Suppose—oh, just suppose, it doesn't cost anything—suppose she could get a close-up today!
>
> The thought warmed her. She had summoned it into mental existence hundreds of times, just for the sake of that warming tingle which came in its wake. After all, it wasn't IMPOSSIBLE! It DOES happen to extra girls—well, not regularly but frequently enough to justify the perennial visualization of its glorious possibility.[56]

But this discourse, which was essential to a romanticized construct of the acting profession that divorced artistry from labor, and glamour from hard work, was also creating a problem of overcrowding and unemployment in Hollywood.

During the spring and summer of 1933, as the number of screen extras skyrocketed and the number of job opportunities shrank, the studios tried to do something about this problem. In one case, the Hays Office warned publicists for the Chicago World's Fair that they would be in trouble if the "Hollywood exhibit [was] found responsible for any addition to the West Coast's unemployment situation." The warning came after the Hays Office discovered that some of the Fair's publicity copy was promising "beautiful Chicago girls and handsome youths" career opportunities in Hollywood. According to the producers' association, "Unofficial screen tests and picture popularity contests, with that free ride to Hollywood, long ago were condemned by the picture industry."[57] The producers' response to the situation, however, may have had more to do with the fact that the World's Fair publicists had acted in an unauthorized capacity. For, as subsequent accounts in *Variety* indicate, the studios had not altogether ceased the practice of holding contests.

In October of 1933, the number of Hollywood potentials increased dramatically when Paramount held its "Panther Woman and Search for Beauty Contests and its much publicized hunt for an 'Alice in Wonderland.' " Problems arose because three hundred of the contest entrants decided to stick around and prove to studio officials that they should have been the winners. In addition, says *Variety*, "Metro [w]asn't help[ing] the stay-away-from-Hollywood cause much by its national tour of a mobile talent testing studio."[58] The major studios justified these measures as part of an organized drive to find new talent that could be developed into star material. They felt new faces were needed to replace the fading box office names and to alleviate the load of current box office stars, whom the studios were overworking in order to protect stories.[59]

The studios' strategy was to sign a large number of people to contracts for three to six months with options, and see if anyone panned out. They figured that "if one or two potential stars [could] be developed from the batch, it would be worth the effort."[60] The studios also revived the talent school system, which provided aspiring actors with a nominal weekly income, some on-the-job training, and, for those who were lucky enough after three months, an opportunity for bit work or feature roles.[61] Many of those who were not chosen to continue, however, remained in Hollywood and tried to get work as extras as they waited for their lucky break from some more observant director or producer. Thus, while studio practices promised to produce a star or two, they otherwise contributed to the growing unemployment situation in Hollywood.

Given the stiff competition for jobs, the opportunities for unknowns were minimal. Statistics based on actor employment figures from May 1932 to April 1933 indicated that of the 9,830 registered principal and bit players hoping to gain a steady livelihood as actors, only 1,102 (nearly a 10-to-1 ratio) had jobs during this period.[62] This ratio was much worse for extras. Of the 18,000 extras registered with the Central Casting Bureau, only an average of 550 were employed by the studios on any given day; most extras were lucky to get eleven or twelve full days of work each year.[63] The influx of new people into Hollywood's extra labor supply also made employment more difficult for existing actors, especially "old-timers" from the silents who had been let go by the majors and relegated to casual employment. In August of

1933, the Academy of Motion Pictures tried to intervene in this situation by compiling and distributing a list of old-timers who were to receive "first call for extra parts," but it could not guarantee employment for these actors.[64]

In the face of the Academy's efforts to alleviate the extra situation in Hollywood, the studios' talent searches and contests were not only wasteful of human resources, they were irresponsible and negligent. Just before the studios stepped up their talent searches, the Academy's Standing Committee on Extras released its report citing the growing unemployment and sweat-shop treatment of screen extras. The findings were presented at the NRA Code hearings and formed the basis of many of the provisions protecting extras' labor in the final NRA Code. During the Academy's investigation, a number of ideas for handling or restructuring the situation also emerged. One plan required "all persons entering Hollywood with the intent of pursuing an extra career" to make their desires known immediately to the Central Casting Bureau. The Bureau would in turn refer each applicant to a "carefully selected talent picking committee." If this committee decided that the applicant demonstrated ample qualifications, he or she would be "officially awarded" the title of performer or extra.[65] This idea had the advantage of funneling people through the star system from the ground floor up and thus, like the talent schools, allowing the studios to maintain better control over the careers of emerging actors. One problem with this system, of course, is that few people came to Hollywood with the dream of becoming an extra. Most of those in the extra ranks ended up there by default, either by failing to achieve a high-ranking status or through downward mobility in the star system.

Contrary to the discourse of stardom found in films such as *Morning Glory*, talent was not necessarily the key ingredient to an actor's success. Hiring and firing decisions were rooted more firmly in economics. During this period, for example, the major studios were signing contract players at an almost unprecedented rate. During one week in September of 1933, more than two hundred actors were signed. *Variety* explained that this situation was due partly to the studios' anticipation of increased production and the need not to be caught waiting for desired players to finish out freelance engagements. The major angle, however, was "the feeling that inflation was just around the corner [and] that would mean upped salaries to meet the new conditions."[66] The recent wave of talent searches and beauty contests shared this same impetus. Though these practices were costly, both in terms of their operation and the aftereffect of unemployment, they were economically beneficial in the long run because they provided studio managers with a means by which to introduce new actors into the star system at lower salaries.

Decisions not to renew silent film actors had likewise been based less on the actors' talent than on the studios' desire to increase box office receipts—in this case, by associating new faces with a new technology. According to an article in *Screen Book*, the aesthetic and technological revolution brought on by the talking picture demanded a new standard of beauty. The old style, epitomized by the "cold immobility" of silent screen stars, was replaced by a modern standard of beauty based on inner warmth, grace of motion, and, above all, personality.[67] Accordingly, this new form of beauty generated an expressiveness and desirability that obviated the need for "talent" in its traditional sense and thus minimized the discursive role of talent in defining what the screen actor should be.

The relationship between talent and beauty nonetheless remained a slippery one that was complicated by differing discourses of stardom perpetuated in Hollywood and on Broadway. In their contests and "talent" searches, for example, the Hollywood studios tended to privilege beauty as a marketable commodity. According to *Variety*,

> It's practically a pushover to get a screen test for a girl who has looks, but for an actress who doesn't appear so hot in person or in a still picture, it remains as tough as ever.
>
> When the agent has a prospect with plenty of talent, but lacking in looks in her street appearance, he has to talk himself out of wind to get attention [from a studio].[68]

Discourses of talent had greater ties to the legitimate stage (as evidenced by *Morning Glory*'s homage to the artistry of great stage actors). By the early 1930s, however, slightly more than two-thirds of the actors under contract with major studios had begun their careers on the stage.[69] This crossover encouraged the integration of star discourses at the same time it provided a form of legitimation for screen acting. Thus, although the studios may have looked initially to beauty for star material, the addition of talent legitimated and gave depth to the spectacle of beauty.

Morning Glory effectively connects these discourses of stage and screen by exploiting the star image of Katharine Hepburn. On one hand, Hepburn's image fit uneasily between the discourses of talent and beauty. According to one account, the question of how to "put Hepburn over" initially gave Hollywood press agents a headache: "She fit into no known pattern for a movie star. But she was too smart to become another glamour girl."[70] As this story implies, however, part of Hepburn's success was due to the way her image combined, yet held in tension, notions of talent and beauty. This tension was put to good use by publicists, who, by separating out different aspects of her image for different venues, could exploit the range of Hepburn's image. A publicity still for *Morning Glory*, for example, represents Hepburn as a Hollywood siren, while the film narrative accentuates her talent. The film further capitalizes on the connection between Hepburn and the character she plays. Hepburn herself had achieved some success in the theater before turning to motion pictures. The film suggests that Eva Lovelace will follow this same trend, because the play she is starring in has motion picture rights attached to it.

By foregrounding discourses of talent in its narrative, *Morning Glory* not only legitimizes the acting profession; it diverts attention away from the conditions of labor in Hollywood. Like *42nd Street*, *Morning Glory* displaces its narratives of stardom onto Broadway. But it also plays down the economic and political aspects of acting by highlighting a romanticized notion of talent and personal sacrifice. The film's brief glimpses into Eva Lovelace's life as an extra, for example, suggest that it is a humiliating, but temporary, period she must endure until she gets her lucky break. The "extra" thus figures discursively as a point of origin in a personalized narrative of upward mobility that hides certain ongoing professional realities for an entire rank of performers. Like the six extras who played the parts of "actors looking for work" during the coffee shop scene of *Morning Glory*, most of these performers remain nameless, receiving no acknowledgment of their work in the film credits. Their history in the acting profession is a fragmented, discontinuous (and often anonymous) narrative that leads nowhere, leaving behind only an accumulation of facts and anecdotes regarding extras' working conditions.

According to the report submitted by the Academy's Standing Committee on Extras, there appeared to be an unwritten policy "to keep wages down to the lowest level and to take advantage of the distress of unemployed players."[71] During the Depression, wages for extras had gone down an average of 20 percent. Whereas an average call netted $9.00 in 1930, it was reduced to $7.48 in 1932, and, by 1933, wages of $2.00 were not unheard of. Many formerly well known actors were working as extras for $3.00 per call.[72] Wages for extras varied also because the roles for extras were not categorized in any consistent manner. An extra named Hugh Lester became so confused by this situation that he wrote a letter to the Screen Actors Guild magazine, *The Screen Player*.

How come I only get seven-five-oh for being a ship's steward, and a lot of other guys get ten for bein' ship's officers? My uniform was just as snappy as theirs was, and God knows I worked just as hard. . . .

A couple of weeks ago, I worked for Paramount. I was a spectator at a night club, in good-looking summer street clothes (Macintosh, forty bucks). My check was for seven-five-oh. Some mugg made a holler and we got a two-fifty adjustment. Now, that was fine, but about four days later I was a pedestrian in the same good-looking summer street clothes and only got a five for my trouble. It was a lot easier bein' a spectator at a night club, even if the floor show was lousy, than it was bein' a pedestrian. How come a spectator rates higher than a pedestrian?

Here's something else I want to know. Why did I get fifteen bucks for bein' a waiter, and four other guys only got seven-five-oh? They made a squawk, but it didn't get them anywhere. I kept my mouth closed, as I figured maybe someone made a mistake and gave me the fifteen thinkin' I was twins.[73]

Production notes for *Morning Glory* indicate this same variance in wages. In one scene, for example, the nine atmosphere players received anywhere between $10.00 and $35.00. Extras playing pedestrians received $7.50 for the day while "actors looking for work" received only $5.00. Actors used for two scenes—a day scene requiring "smart winter clothes" and a night scene requiring "full formal evening dress"—received $10.00. Yet other actors who were used for only one scene requiring formal evening dress received between $10.00 and $25.00.

Extras in Hollywood were subject to a number of abuses as well. In one case, a director "forced an extra player to turn his back to the camera while the director spoke lines into the mike that should have been said by the extra player."[74] By doing so he saved his studio the difference between $7.50 and $25.00. In another case, extras hired to play a crowd scene were asked to wear a hat and topcoat and to bring along a second pair of these items. When they arrived on the set they were ordered to put the extra clothing on dummies to help fill the background. No additional wages were given for these efforts.[75] By 1935 the Screen Actors Guild was so distressed by the treatment of extras that it ran a feature in its magazine titled "Are Extras People?" The story recounts some rather serious physical abuse of extras:

Forty women received the call to report at 5:30 P.M. to the set for "Riff Raff" which J. Walter Ruben was directing for Metro-Goldwyn-Mayer. The call from Central Casting Corp. had specified "light rain." . . . The set worked twelve hours until 5:30 A.M., and each woman received a check-and-a-half—$11.25. Lunch was not called until twelve midnight—six-and-one-half hours after the extras reported to the set.

A few minutes after 10:00 P.M., the women were ordered into the rain for the first time. The set was equipped with overhead sprinklers, three fire hoses, and three wind machines. The latter created such a terrific gale that a number of women were forcibly knocked down and bruised in each take. One woman was knocked unconscious while another who took the full force of the stream of water from the hose on her back, was paralyzed from her hips down for several hours. Four women were temporarily blinded when the water hit them full force in the eyes.[76]

According to the SAG, no law prevented such abuse of actors. Although the California law restricting the employment of women to eight hours a day might have otherwise averted the foregoing situation, studios found loopholes that allowed them to sidestep the law. A portion of the Industrial Welfare Commission Order No. 16-A stated:

> No employer shall employ or suffer or permit any woman extra receiving a wage of $15.00 or under per day or a wage of $65.00 or under per week to be employed more than eight hours in any one day of twenty-four hours, except that *in the case of emergency* women may be employed in excess of eight hours.[77]

Studios sidestepped the financial provision of the law by paying just over the legal limit; a common wage for chorus girls, for example, was $66.00 per week. They also manipulated the provision that allowed for extras to work overtime. According to the SAG, the problem with Order No. 16-A arose because it did not define the term "emergency." Thus, when studios were running behind schedule on their film production, they simply called an emergency. During a ten-month period in 1935, the Guild learned that there had been an "emergency" almost every day at one or more of the major studios, causing anywhere between five women extras and two hundred women extras to work from two to eight hours overtime (see Table 1).

The SAG had been researching and monitoring the working conditions of actors since its inception, but it was often the job of actors on the set to police the day-to-day actions of producers. Aside from demanding overtime pay and comparable wages for comparable work, extras balked against the use of nonprofessional actors for extra roles. During the filming of *College Holi-*

Table 1. Women extras overtime on major studio sets
(January 1, 1935, to October 15, 1935)

Studio	No. cases	¼ checks	½ checks	¾ checks	Full checks	No. directors
Columbia Pictures Corp.	12	8	4	. . .	. . .	6
Fox Film Corp. Studio[a]	40	21	14	4	b	13
Metro-Goldwyn-Mayer Studio	49	41	3	5	. . .	18
Paramount Studio	28	17	10	1	. . .	10
RKO-Radio Pictures Studio	49	29	10	3	c, d	20
United Artists Studio[e]	11	7	1	2	1	4
Universal Studios	16	13	. . .	2	. . .	6
Warner Bros.-First National Studio	50	33	12	3	2	15

[a] Includes one Twentieth Century-Fox.
[b] For 17 hours, extras received one ¾ check. Another ¼ check was received by each woman, as an adjustment.
[c] Eighty-five women forced to remain on the set for 17 hours.
[d] Under director Stevens for six days from September 9 to September 14, extras received from ¼ to ¾ checks every day.
[e] Includes Samuel Goldwyn, Inc., Edward Small Productions, and Twentieth Century Pictures.
The above figures are based on fewer than 12 women members of the Junior Guild.

day in Santa Barbara, for example, a group of women extras put a stop to the hiring of rich debutantes who wanted the "thrill" of working in pictures: "Get rid of those society girls and give those jobs to legitimate extras, or we'll walk out—and every star in the cast goes with us!" they warned.[78] Since their demand was backed by Jack Benny, George Burns, and Gracie Allen, the studios relented and replaced the local thrill seekers with professionals.

Much of the available information concerning abuses of actors (especially extras) was recorded after 1933—after the formation of the Screen Actors Guild and after the passage of the movie industry's NRA Code of Fair Competition. NRA labor policies, in other words, did not prevent abuses from occurring. Studios violated these policies whenever it suited them, and, until 1935, they even ignored the Guild's legal right to represent actors in collective bargaining situations. In their hands the NRA thus became a discursive tool that effectively shielded the public from these facts. The studios' goodwill advertising, propaganda films, and overall public relations efforts created an image of industrial unity and national concern. The feature-length entertainment films additionally provided studios with a means to rewrite labor–management conflict in Hollywood *(42nd Street)* or to foreground discourses of stardom that diverted attention away from such conflict *(Morning Glory)*.

Morning Glory achieved widespread popularity, grossing $65,000 in only the first four days of its release.[79] Audience attention was focused especially on Katharine Hepburn, a rising star at RKO who would later win the Academy Award for best actress for her portrayal of Eva Lovelace. It is perhaps fitting that in the most tumultuous year of actor–producer relations in Hollywood's history, the Academy would present one of its prestigious awards to someone for her portrayal of an actor who characterized the discourse of entertainment and ideology of stardom that was so profitable to studio management. Although there were members within the Academy who were concerned about protecting actors through such means as the Standing Committee on Extras, the Academy was still an organization that was controlled by producers and that functioned as a powerful discursive machine on their behalf. As tokens of the movie industry's appreciation of film excellence, the Academy Awards seem harmless enough. But they stand as a symbolic gesture of the producers' vested interest in diverting our attention away from the struggles that have constituted the institution of stardom.

6 / The Terrain of Actors' Labor

"Doing cultural studies," says Lawrence Grossberg, "is not a matter of merely continuing the work that has already been done, staying on the same terrain, but of asking what is left off the agenda in relation to specific contexts and projects."[1] Thus, while my attempt to theorize actors' subjectivity and to historicize actor–producer relations in the U.S. film industry might be said to fill a gap in previous studies or to chart a new terrain, my project is also meant to resituate star studies in relation to cultural studies by interrogating the historical and theoretical boundaries of cultural studies itself. What I have discovered in the process is that a new understanding of the interrelationships among labor, subjectivity, discourse, and history is necessary if cultural studies is to account for the subjective role of film actors within the cinematic institution.

Theorizing labor and reestablishing labor at the heart of cultural studies was most obviously the key to these investigations. The concept of labor first had to be rescued from a post-Marxist intervention that rejected the crude economism of Marx's base-superstructure model but, in the process, privileged a relatively autonomous notion of culture that held little place for the laboring subject. In other words, in its eagerness to dissociate cultural studies from an economic determinism, the post-Marxist project relegated the notion of labor to the industrial base, where it was all but forgotten. How can we account for this ironic twist?

I believe the main reason the concept of labor has never been central to cultural studies is that cultural theorists have failed to see how labor can be analyzed discursively. Tied to the concept of wage, and perceived as the product of a worker's effort to be bought or exchanged in the marketplace, labor never took on a discursive dimension. Although Raymond Williams argued for a broader concept of labor that would, among other things, account for the production of language, he never placed a theory of labor at the center of his investigative projects. Other cultural theorists did not develop the theoretical possibilities that Williams raised. Labor has thus remained a blind spot for cultural theorists—what Williams might call a "residual form" in the process of rearticulating Marxism for cultural studies. As the last stubborn outpost of orthodox theory, this blind spot creates a "labor reductionism" not unlike the problem of class reductionism that cultural theorists have already recognized and critiqued.

The discursive dimensions of labor and the instability of labor as a fixed characteristic of cultural production become clearer when labor is examined in relation to subjectivity. For if a subject is always a "subject in discourse," then the labor role performed by that subject is discursively determined. The case of screen actors throws this idea into high relief. As subjects caught between their positions as laborers and commodity images, they are involved in a conflict over the very terms of representation. How actors' labor is defined, and who defines it, are issues that delimit an arena of struggle over the material and discursive aspects of subject identity and make the interrelationships among self-representation, filmic representation, and studio representation a scene of constant negotiation.

The discursive aspects of actors' labor and subjectivity make the task of writing history rather messy. Add to this the cultural studies imperatives to break up the linear, cause-and-effect model of traditional history and to attempt to understand the *breadth* of social relations through an examination of the multiple, interrelated practices involved in the historical process, and the task becomes nearly impossible. This is why I have tried whenever possible to restrict my analysis of labor–management relations and actors' (self-)representation to the year 1933—or, even more specifically, to an eight-month period beginning in March 1933 and ending in October 1933. By articulating the various material and discursive practices involved in social relations during this brief period, I have tried to establish the extent to which the struggle over actors' subject identities entered into all aspects of the cinema's production-exchange process. Although historical periodization cannot contain these struggles, it does allow for a clearer articulation of the competing and contradictory discourses that circulate within and around particular events and practices.

As Grossberg argues, the practice of articulation "is both the practice of history and its critical reconstruction, displacement and renewal."[2] It reworks the historical context into which practices are inserted by emphasizing the spatial over the temporal, the possibilities over the certainties, and the discontinuities as well as the linkages between social relations and practices. In mapping the war of position over actors' labor and subjectivity, the practice of articulation has led to encounters with "real historical individuals and groups, sometimes consciously, sometimes unconsciously or unintentionally, sometimes by their activity, sometimes their inactivity, sometimes victoriously, sometimes with disastrous consequences, and sometimes with no visible results."[3]

In the final analysis, no single narrative or combination of narratives will suffice as "the history" of actor–producer relations and the discourses of labor that either arise out of or determine those relations. More likely, the analyses I provide in this book simply point to the many "histories" yet to be written on this topic. I have said very little, for example, about the relationship that actors have had to their agents or to other workers in the film industry. Since I have restricted my analyses to a specific time period and political context, there is also much work to be done on the construction of actors' labor both within different historical moments (e.g., the McCarthy era or the transition to television) and within different national film industries. It would be interesting as well to see how a theory of actors' labor and subjectivity could inform the analyses of other films that comment upon actors' professional and private lives (e.g., *A Star Is Born*, *Mommie Dearest*).

In this final chapter I would like to map out some additional possibilities for the analysis of actors' labor. First, however, I wish to examine how the theoretical perspective of this research contributes to and challenges the current field of star studies. As noted previously, the majority of work in this field analyzes the actor as a "star image" or "text" and thus reduces the actor to the status of object. While concerned with the ideological effect that star images have on spectators, this approach tends to accept the actor/star as a preconstituted phenomenon of industrial discourse. As Barry King notes, "Such a form of analysis, while not necessarily invalid, tends to preclude a consideration of how the discourses of stardom are constructed and activated in the first place."[4] In addition, this approach often neglects consideration of the actor as a social agent or political subject who actively participates in or resists studio discourses of stardom and the material conditions upon which they rest. This is why I have addressed actors as workers, examining the social relations of power that determine and influence their status as subjects in the industry, and why I have avoided a fetishized reduction of actors as objects within the representational field.

Might the term "actor as worker" itself be construed as a Marxist fetishization of the actor? Perhaps. But this discursive construct at least establishes a political perspective that is not complicit with managerial discourses of actors and that, as a result, permits actors' voices to be heard. More importantly, when this construct is defined in relation to subjectivity, it releases the actor from his or her object status without having to resort to such constructs as "personality" or "real person." As Richard Dyer notes, "The whole media construction of stars encourages us to think in terms of 're-ally.' "[5] But the "actor as worker" (understood as a social subject who works and is positioned within the acting profession) provides a space for interrogating what it means to be constituted as an actor. The "actor as worker" thus serves as a theoretical starting point from which to analyze the material and discursive struggles over actors' subjectivity. As a challenge to the elitism of text-based analyses concerned only with stars, this construct also serves to question the appropriateness of the label "star studies" in describing this area of research. (I would prefer, in other words, the more neutral term "actor studies." But until a consensus on this alternative is reached, I will continue to use the familiar term.)

Understanding subjectivity from the perspective of labor has even further implications. It dethrones the spectator as the privileged subject of film studies, and it potentially offers new ways to understand the process of spectatorship. The issue of labor, for example, poses a challenge to those star studies that draw upon subcultural theories of reception to situate meaning—or the power to make meaning—at the level of the audience. According to Dyer, the responsiveness of subcultural groups to stars stems from such groups' exclusion from adult, white, male, heterosexual culture; they identify with and appropriate star images that they perceive as representative of this exclusion.[6] In *Heavenly Bodies*, Dyer analyzes how gay reception of the Judy Garland image is rooted in an identification with her as a tragic and exploited, yet enduring and dignified, figure.[7] In his now somewhat infamous analysis of Madonna, John Fiske argues that the star's image allows young girls to "struggle against the patriarchy inscribed in them."[8] Subsequent studies of Madonna, such as those found in *The Madonna Connection*, demonstrate the range of subcultural readings that can be derived from a single star image.[9]

While work in this area has flourished over the last decade, it has not gone without critique. King, for example, argues that "Dyer's view of stardom amounts to a role-conflict theory, with certain categories of audience member—such as adolescents, blacks, women and gays—seen as particularly responsive to the contradictory play of meaning and identity implicit in the imagery of stars."[10] One's interpretation of a star image or filmic text, in other words, is somehow linked to the condition of being young, gay, black, or female. Mike Budd, Robert M. Entman, and Clay Steinman launch a more general attack on what they perceive as the "affirmative character" in the cultural studies analyses of Fiske and others:

> Cultural studies tends to affirm that people habitually use the content of dominant media against itself, to empower themselves. [This suggests that] we don't need to worry about people watching several hours of TV a day, consuming its images, ads, and values. People are already critical, active viewers and listeners, not cultural dopes manipulated by the media.[11]

Though this latter point provides a needed word of caution, it remains important to understand how marginal groups *do* manipulate the popular media for their own purposes (even while we recognize that all spectators—regardless of race, class, gender, or sexual orientation—read media texts in contradictory and appropriative ways).

Putting questions of identity and marginality aside for the moment, what we might consider in cases of (subcultural) reception is the labor performed by the spectating subject. Spectators, in other words, are "film workers" who perform specific labors in relation to the cinema. The amount or type of labor that spectators perform is determined by the way they are positioned or position themselves in relation to the cinema and other social practices. Seen in this light, a spectator's appropriation of or resistance to meaning is not a behavioral reaction caused by a certain "condition of being," but a form of labor that becomes necessary in order to make sense of (or derive pleasure from) a particular subject position. Thus, gay and lesbian reception can be viewed in terms of the labor involved in resisting heterosexist narratives or in reformulating "straight" representation into camp readings. Gay and lesbian spectators must also expend more physical effort in locating gay films and videos since these do not enjoy the same degree of visibility and circulation as their mainstream counterparts. The notion of a spectator's labor must therefore extend *beyond* the concept of reading practices.[12] For, in addition to constructing meanings out of film texts, spectators perform a variety of activities (ranging from renting videos to having conversations about the films they see) that form an integral part of the viewing process. These "labor practices" arise out of the relation between labor power and spectatorial identities. What we do with films, who does what to them, is finally a matter of labor power—something that potentially can be mapped according to a theory of labor power differences.

By highlighting the possibilities of the spectator's laboring practices I do not mean to trivialize the labor performed by actors in an industrialized setting. Actors' labor is, after all, *paid* labor, and conditions of production and consumption will necessarily result in varying types and intensities of labor practices. Nonetheless, when the relationship between cinema and subjectivity is approached from the perspective of labor, our understanding of the relationship between actors and spectators and of their potential commonalities changes as well. King's work has made some inroads here by suggesting that the actor–spectator relationship is rooted in the conditions of capitalism. The star, he argues, "can function as a metonymy for labour power and for what stands behind this connection—that is, the sensuous creative capacity of human labour power."[13] Dyer also alludes to this notion in *Heavenly Bodies*. At one level, he states, stars "articulate a dominant experience of work itself under capitalism—not only in the sense of being a cog in an industrial machine, but also the fact that one's labour and what it produces seem so divorced from each other."[14] The implication is that spectators and actors experience a common alienation from the labor process as workers within a capitalist economy. Accordingly, spectators may form an unconscious identification with star images based on labor alienation (as opposed, for example, to the oedipal scenario suggested by psychoanalytic theorists). The concepts of labor alienation and identification, however, should not be seen in economically reductive terms, that is, as products determined by the capitalist mode of production. As I have stressed throughout this book, labor and labor power differences are constructed both materially and *discursively* within social practices. Spectatorial forms of identification with actors are thus influenced by the process by which labor power connects with subject positionalities.

On a practical level, this means that spectators identify with the labor practices actors perform in a number of cultural and institutional spheres. For, as King notes, "identification with stars often takes the form of an identification with a specific form of *agency*—with the star per se rather than with the narrative functions (characters) he or she represents."[15] This emphasis on agency tends to collapse not only the traditional "star image" versus "real person" dichotomy, but also the "star" versus "viewer" dichotomy. Consequently, as social subjects who must navigate the gendered, racialized, and otherwise politicized space of cultural practices, actors and spectators come to occupy similar positions and to experience similar struggles in terms of labor power and subject identity. Spectators may not experience the public visibility of stars, but they can identify with the physical and emotional work it takes to campaign on behalf of abortion rights (Lauren Bacall), the rights of farm workers (Jessica Lange), gay and lesbian rights (Lily Tomlin), the preservation of the Amazon rain forest (Madonna and others), or the problem of sexual abuse (Oprah Winfrey). In this way, actors and spectators form alliances that are vastly different from the narrow star (as object)–viewer (as subject) relation posited by mainstream film studies. Here the star/actor signifies a positionality and rallying point for matters of cultural resistance that are inflected by, but go beyond, the industrial context of Hollywood entertainment.

These ideas regarding the spectator and the viewer–spectator relationship are provisional at best, but they should indicate a few of the ways in which a theory of labor might challenge or revitalize current notions of subjectivity in the cinema. In terms of rethinking the issue of actors' labor and social agency, additional possibilities lie in the realm of national policy making. For example, in *The Power and the Glitter*, journalist Ronald Brownstein traces the Hollywood–Washington connection back to the 1920s. He notes that the major change that has occurred since that time is "the growing sense in Hollywood of its own political legitimacy."[16] Whereas actors initially functioned as passive stage ornaments for political campaigns, they are now active participants in a range of political arenas. The Creative Coalition (TCC), headed by actor Ron Silver, is one current example of an advocacy group designed to "combat the image of the Hollywood dilettante." Boasting "state-of-the-art celebrity activism," its members make it a point to become informed on the details of the causes that involve them. The three-hundred-plus membership has tackled issues such as national health care, the environment, and AIDS. It has also been active in the fight over funding from the National Endowment for the Arts.[17]

Speculations about the motivations behind actors' political activism are numerous. Some charge actors with egotism; others see it as a form of altruism, a means to achieve validation, or a way to redeem the reputation of the acting profession. Brownstein suggests further that it provides a way for actors to "control their own celebrity."[18] Such speculations, however, are not useful in and of themselves. Specific attempts at activism must be placed within their historical context and analyzed in terms of the struggles that occur over labor power and subjectivity. By examining the ways in which actors mobilize politically or are positioned by the political process, we can learn more about the way labor power differences construct actor–spectator relations and actor–government relations around issues of economic, political, and cultural significance.

Because the issue of employment continues to be a primary concern for actors, we also need to examine actors' labor in a more literal sense. How are actors' jobs and working conditions affected by the changing nature of the film industry? How does the "actor as worker" get constructed discursively in various production venues? One of the film industry's responses to the faltering state of the current U.S. economy, for example, is to shoot films in midsized cities where the cost of living is relatively low and producers can avoid hiring union personnel. This has led one group of actors in Pittsburgh to form an organization called Pittsburgh Actors Seeking Answers (PASA). Although the film industry pumped an estimated $63 million into Pittsburgh's local economy in 1991, local actors working as screen extras made just $25.00 per day (only $15.00 to $20.00 more than screen extras made per day in 1933!). One of the problems, says PASA, is that one casting agency has a monopoly on the casting market and consistently underbids the rest. This places local actors in a double bind. Though grateful for the work, actors are affected adversely by the industry's trickle-down economics. The stated goal of PASA is to "seek equity for qualified professionals." In the process it exposes the labor power differences among various groups and businesses in the film industry. In a letter to the *Cityflickers*, a Pittsburgh newspaper that reports on the local film scene, PASA member Louis Spenser poses the question, "How much of the $63 million did *you* see in 1991?"[19]

A comparison between actors' working conditions in the 1930s and the 1990s indicates that some things have changed very little. In addition to the continued exploitation of the industry's "casual labor" force, problems exist in the area of work classification and entitlement. A 1933 article in *Variety*, for example, addresses the plight of Hollywood "stand-ins" who were "engaged to take the star's place while the lighting and cameras were arranged . . . only to step out of the picture when the real scenes [were] to be photographed."[20] While these stand-ins were paid somewhat better than screen extras, they never received screen credit, and few of them made it very far in motion pictures. Once branded as stand-ins, they often became the personal maids or valets of established stars who kept them on their payrolls between pictures.

A similar situation exists for today's body doubles, actors who function not merely as stand-ins (in the sense just described) but who are filmed and included as part of the motion picture. Although body doubles have been used in Hollywood for some time, their current situation is exacerbated by the increasingly explicit use of nudity and sexuality in film. As one Hollywood publicist notes, "It's standard practice to have a body double present in case an actress feels uncomfortable doing a scene."[21] But Shawn Eileen Lusader, a body double for Anne Archer, wants to know, "Why is it acting when Anne does it and stripping when I do it? . . . She gets to enhance her career using my body."[22] In response, Lusader has formed an organization called the Body of Doubles Committee (BOD). Body doubles, she explains, have little control over the use of their likenesses in films and publicity, receive no residuals for their likenesses from videos and overseas sales, and often encounter sexual harassment during auditions and film production. But, in addition to seeking better compensation and improved working conditions, BOD is fighting for screen credit and the reclassification of body doubles from extras to principal players.

The issue of screen credit is one of the major obstacles faced by BOD. Producer Steven Deutsch, who argues against screen credit, explains, "You're trying to create an illusion and you don't want to tell the audience, 'You've just seen an illusion.' "[23] But Lusader points out that stunt doubles are listed in the credits, and audiences do not seem deterred in the face of this illusion-breaking phenomenon. The matter at present is being considered by the Screen Actors Guild. According to Mark Locher, a spokesperson for SAG, stunt doubles did not always get screen credit. It was only through organizing that they achieved recognition and better compensation.[24] Perhaps body doubles can learn from this history of organizing and achieve better working conditions for actors who labor in this capacity.

The lesson to be learned by film scholars is that there is a great deal more going on behind the "scenes" of star images. Actors are laboring subjects who encounter and must negotiate the ongoing economic, political, and discursive practices of their profession within the film industry. From a contemporary perspective, film scholars need to address the changing political economy of the film industry and how this affects the role of actors and other film workers. In addition to analyzing the diversity of production sites, we need a better sense of the relations among actors within the profession's hierarchy; of how various material and discursive conditions affect their current struggles over defining subject identities; and of how the interrelationships among different entertainment media influence the construction of the acting profession. Although cultural studies theorists have thus far reserved ethnographic analysis for spectators, ethnographic techniques would undoubtedly benefit research in this area. This is especially so in the case of screen extras since information regarding their labor conditions—and, more importantly, their responses to these labor conditions—tends to go undocumented.

The research possibilities I mention here will take us into new, uncharted areas of star studies and cultural studies. I do not want to leave the impression, however, that these suggestions are without precedent. Barry King's work clearly has been formative in developing a theoretical framework that addresses actors as laborers. Although I have been critical of Richard Dyer's work throughout this book, he almost singlehandly developed the field of star studies as we currently know it. No one working in this field can fail to acknowledge a debt to the theoretical and ideological groundwork he laid. While his work is admittedly text-based, he occasionally marks the importance of actors' role as laborers. In *Heavenly Bodies*, for example, Dyer notes that the three stars he examines (Marilyn Monroe, Paul Robeson, and Judy Garland) all "revolted against the lack of control they felt they had."[25] The way his analysis of Robeson addresses struggles over subject identity has also been helpful to me in developing my theoretical approach to actors' labor and subjectivity. I am indebted finally to the spirit of cultural studies, which not only provides a space for, but insists upon, constant challenges to previous theoretical work. Just as cultural theorists have exploded the way we think about texts, or about the conditions of reception, I hope to challenge the way we think about actors and their relation to labor, representation, and subjectivity.

I began this project by raising the specter of "lack." Perhaps a better tactic would be to talk about direction, for it seems to me that the field of star studies has stalled in recent years. Textual analyses of star images travel a well-worn path, and subcultural reception studies of stars tend to reveal more about the identity formations of social groups than about the role of actors. While a cultural studies perspective has been used by film scholars to analyze various aspects of film practice, its lack of an adequate theory of labor has prevented scholars from addressing the "subjects of production." Through the development of a theoretical perspective that allows cultural studies to account for the "actor as worker," the field of star studies (and film studies in general) can move in a new direction.

This theoretical framework, first and foremost, suggests that (post-) Marxist formulations of labor still have a great deal to offer film scholars, and can be applied in ways that are nonreductive. In terms of historiography, an analysis of labor struggles not only adds a new dimension to history, but problematizes the way we perceive the social relations of film. Attention to the competing and interrelated aspects of material and discursive (labor) practices furthermore allows for an understanding of how these social relations affect subjectivity. Finally, analyses based on labor power differences can begin to inform the social relations between film workers and spectators; the specific labor(s) performed by viewing subjects; and the forms of resistance experienced by subjects at various points in the cinema's production-exchange process.

Notes

Preface

1. By restricting my discussion to *Screen* theory and cultural studies I do not intend to minimize the influences and impact of feminist theory, postcolonialist theory, and other theoretical trends. I merely want to highlight the visible differences between these two general approaches to film and media study.

2. Richard Dyer, *Stars* (London: British Film Institute, 1979).

3. See Richard deCordova, "The Emergence of the Star System in America," *Wide Angle* 6.4 (1985): 4-13; *Picture Personalities: The Emergence of the Star System in America* (Urbana: University of Illinois Press, 1990); and Barry King, "Articulating Stardom," *Screen* 26.5 (1985): 27-50; "The Star and the Commodity: Notes Towards a Performance Theory of Stardom," *Cultural Studies* 1.2 (1987): 145-61; "Stardom as an Occupation," in *The Hollywood Film Industry*, ed. Paul Kerr (London: Routledge, 1986): 154-84.

4. James Naremore, *Acting in the Cinema* (Berkeley: University of California Press, 1988).

5. David Bordwell, Janet Staiger, and Kristin Thompson, *The Classical Hollywood Cinema: Film Study and Mode of Production to 1960* (London: Routledge & Kegan Paul, 1985).

6. DeCordova's *Picture Personalities* would be a notable exception here.

1 / The Actor's (Absent) Role in Film Studies

1. Richard Dyer, *Stars* (London: British Film Institute, 1979).

2. Christine Gledhill, ed., introduction to *Star Signs* (London: British Film Institute, 1982): iv.

3. See James Naremore, *Acting in the Cinema* (Berkeley: University of California Press, 1988): 9-21.

4. Jeremy G. Butler, ed., introduction to *Star Texts* (Detroit: Wayne State University Press, 1991): 7.

5. Richard deCordova, *Picture Personalities: The Emergence of the Star System in America* (Urbana: University of Illinois Press, 1990): 8.

6. Ibid., 18-19.

7. Barry King, "The Star and the Commodity: Notes Towards a Performance Theory of Stardom," *Cultural Studies* 1.2 (1987): 145, 149.

8. See also Barry King, "Articulating Stardom," *Screen* 26.5 (1985): 27-50; and "Stardom as an Occupation," in *The Hollywood Film Industry*, ed. Paul Kerr (London: Routledge, 1986): 154-84.

9. Vincent Mosco, "Critical Research and the Role of Labor," *Journal of Communication* 33.3 (1983): 237.

10. Richard deCordova, "The Emergence of the Star System in America," *Wide Angle* 6.4 (1985): 5.

11. Ien Ang has noted a similar situation within television studies, arguing that "our knowledge about television audiencehood has been colonized by [an] institutional point of view" that symbolically constructs the "audience" as an objectified category of others that can be controlled and constrained. See Ang, *Desperately Seeking the Audience* (New York: Routledge, 1991): 2, 7.

12. Gorham Kindem, "Hollywood's Movie Star System: A Historical Overview," in *The American Movie Industry*, ed. Gorham Kindem (Carbondale: Southern Illinois Press, 1982): 93.

13. Richard Maltby, " 'Baby Face' or How Joe Breen Made Barbara Stanwyck Atone for Causing the Wall Street Crash," *Screen* 27.2 (1986): 26.

14. Murray Ross, *Stars and Strikes* (1941; reprint, New York: AMS Press, 1967).

15. David F. Prindle, *The Politics of Glamour: Ideology and Democracy in the Screen Actors Guild* (Madison: University of Wisconsin Press, 1988).

16. Larry Ceplair and Steven Englund, *The Inquisition in Hollywood* (Berkeley: University of California Press, 1979); Michael Nielsen, "Toward a Workers' History of the U.S. Film Industry," in *The Critical Communications Review, Volume I: Labor, the Working Class, and the Media*, ed. Vincent Mosco and Janet Wasko (Norwood, N.J.: Ablex, 1983): 47-83. See also Samuel Lipkowitz, "Collective Bargaining in Motion Picture Production" (master's thesis, American University, 1939).

17. Thomas Schatz, " 'A Triumph of Bitchery': Warner Bros., Bette Davis and *Jezebel*," *Wide Angle* 10.1 (1988): 17. See also Alexander Doty, "The Cabinet of Lucy Ricardo: Lucille Ball's Star Image," *Cinema Journal* 29.4 (1990): 3-22; Charles Eckert, "Shirley Temple and the House of Rockefeller," *Jump Cut* 2 (1974): 1, 17-20; Florence Jacobowitz, "Joan Bennett: Images of Femininity in Conflict," *CineAction!* 7 (1986): 22-34; Richard Lippe, "Kim Novak: A Resistance to Definition," *CineAction!* 7 (1986): 4-21.

18. Jane Gaines, *Contested Culture: The Image, the Voice, and the Law* (Chapel Hill: University of North Carolina Press, 1991).

19. See Walter Benjamin, "The Work of Art in the Age of Mechanical Reproduction," in *Illuminations*, ed. Hannah Arendt (New York: Schocken Books, 1969).

20. Hugo Munsterberg, *The Photoplay: A Psychological Study* (New York: Appleton, 1916): 178.

21. André Bazin, *What Is Cinema?* trans. Hugh Gray (Berkeley: University of California Press, 1967): 102.

22. Maurice Yacowar, "An Aesthetic Defense of the Star System in Films," *Quarterly Review of Film Studies* 4.1 (1979): 41-42.

23. Ibid., 39.

24. James Naremore, *Acting in the Cinema* (Berkeley: University of California Press, 1988): 26.

25. Charles Affron, *Star Acting* (New York: Dutton, 1977): 3.

26. Naremore, *Acting in the Cinema*; Carole Zucker, ed., *Making Visible the Invisible* (Metuchen, N.J.: Scarecrow Press, 1990).

27. Affron, *Star Acting*, 4.

28. See Laleen Jayamanne, "Modes of Performing: Bodies & Texts (some thoughts on fe/male performances)," *Australian Journal of Screen Theory* 9/10 (1981): 124.

29. Affron, *Star Acting*, 5.

30. Naremore, *Acting in the Cinema*, 30.

31. Dyer, *Stars*, 72.

32. Pam Cook, "Star Signs," *Screen* 20. 3/4 (1979/80): 87.

33. King, "The Star and the Commodity," 148.

34. Richard Dyer, *Heavenly Bodies* (New York: St. Martin's Press, 1986): 5.

35. Gledhill, *Star Signs*, ix.

36. Anne Friedberg, "Identification and the Star," in *Star Signs*, 50.

37. See, for example, Andrew Britton, *Katharine Hepburn* (Newcastle upon Tyne: Tyneside Press, 1984); Jacobowitz, "Joan Bennett"; Lippe, "Kim Novak"; and Simon Watney, "Katharine Hepburn and the Cinema of Chastisement," *Screen* 26.5 (1985): 52-62.

38. Pam Cook, "Stars and Politics," in *Star Signs*, 26.

39. Dave Morley, "Text, Readers, Subjects," in *Culture, Media, Language*, ed. Stuart Hall, Dorothy Hobson, Andrew Lowe, and Paul Willis (London: Hutchinson, 1984): 172.

40. Stuart Hall, "The Rediscovery of 'Ideology': Return of the Repressed in Media Studies," in *Culture, Society and the Media*, ed. Michael Gurevitch, Tony Bennett, James Curran, and Janet Woollacott (London: Hutchinson, 1982): 85.

41. Terry Lovell, *Pictures of Reality* (London: British Film Institute, 1983): 43.

42. Chantal Mouffe, "Hegemony and New Political Subjects: Toward a New Concept of Democracy," *Marxism and the Interpretation of Culture*, ed. Cary Nelson and Lawrence Grossberg (Urbana: University of Illinois Press, 1988): 91.

43. See Nicholas Garnham, "Subjectivity, Ideology, Class and Historical Materialism," *Screen* 20:1 (1979): 126.

44. Raymond Williams, *Marxism and Literature* (New York: Oxford University Press, 1977): 91.

45. Ibid., 97.

46. Stuart Hall, "Culture, the Media and the 'Ideological Effect,' " in *Mass Communication and Society*, ed. Michael Gurevitch, James Curran, and Janet Woollacott (London: Sage, 1979): 316.

47. Hall, "The Rediscovery of Ideology," 84.

48. Karl Marx, *Wage-Labour and Capital* and *Value, Price and Profit* (New York: International Publishers, 1988): 38. See also Karl Marx, *Capital*, vol. 1, trans. Ben Fowkes (New York: Vintage, 1977).

49. Martyn J. Lee, *Consumer Culture Reborn* (London: Routledge, 1993): 6.

50. Marx, *Wage-Labour and Capital*, 39.

51. Lawrence Grossberg, *We Gotta Get Out of This Place* (New York: Routledge, 1992): 55.

52. Ibid., 54.

53. For a discussion of the theory–history problematic in cultural studies, see Graeme Turner, *British Cultural Studies* (Boston: Unwin Hyman, 1990): 180-88.

54. Richard Johnson, "What Is Cultural Studies Anyway?" *Social Text* 6.1 (1987): 43.

2 / The Subject of Acting

1. See, for example, Stuart Hall, "Culture, the Media and the 'Ideological Effect,' " *Mass Communication and Society*, ed. James Curran, Michael Gurevitch, and Janet Woollacott (London: Sage, 1979): 336-42.

2. Jane Gaines, "In the Service of Ideology: How Betty Grable's Legs Won the War," *Film Reader* 5 (1982): 56. Gaines discusses at length the relation between the star image and Marx's theory of commodity fetishism.

3. Marian L. Mel, *Method of Employment of Extra Players in the Motion Picture Industry in California* (1930): 1, as quoted in Murray Ross, *Stars and Strikes* (1941; reprint, New York: AMS Press, 1967): 77.

4. Numerous instances of abuse are documented in the actors' union journal *Screen Player*. Also see Ross, *Stars and Strikes*, pp. 120-26.

5. Hortense Powdermaker, *Hollywood, the Dream Factory* (Boston: Little, Brown, 1950): 207.

6. The term "isolation effect" is from Nicos Poulantzas, *Political Power and Social Classes* (London: New Left Books, 1973).

7. "Actors' Chances Put at 10 to 1," *Variety* (9 May 1933): 7. The report states that at this time more than 10,000 persons, exclusive of extras, were hoping to gain a livelihood as actors.

8. Alexander Walker, *Stardom: The Hollywood Phenomenon* (New York: Stein & Day, 1970): 259.

9. Gaines, "In the Service of Ideology," 53.

10. Tony Bennett, "Theories of the Media, Theories of Society," in *Culture, Society and the Media*, ed. Michael Gurevitch, Tony Bennett, James Curran, and Janet Woollacott (London: Hutchinson, 1982): 52.

11. Walker, *Stardom*, 262.

12. Barry King, "Articulating Stardom," *Screen* 26.5 (1985): 46–47.

13. Ibid., 45.

14. Ibid.

15. See Walker, *Stardom*, 260.

16. King, "Articulating Stardom," 45.

17. Barry King, "Stardom as an Occupation," in *The Hollywood Film Industry*, ed. Paul Kerr (London: Routledge & Kegan Paul, 1986): 168.

18. Screen Actors Guild report as quoted in Powdermaker, *Hollywood, the Dream Factory*, 210.

19. See, for example, Thomas Schatz, " 'A Triumph of Bitchery': Warner Bros., Bette Davis and *Jezebel*," *Wide Angle* 10.1 (1988): 16–29.

20. Quoted from Ginger Rogers's contract with Warner Bros., 1933. This clause was representative of many star contracts during this era.

21. Jane Gaines, *Contested Culture: The Image, the Voice, and the Law* (Chapel Hill: University of North Carolina Press, 1991): 160.

22. King, "Stardom as an Occupation," 168.

23. Gaines, *Contested Culture*, 149.

24. Ibid., 154.

25. John Ellis, "Star/Industry/Image," in *Star Signs*, ed. Christine Gledhill (London: British Film Institute, 1982): 3.

26. Ibid.; Tony Bennett and Janet Woollacott, *Bond and Beyond* (New York: Methuen, 1987): 271.

27. "Hungry Need for New Film Faces, Every Studio on Talent Hunt," *Variety* (10 October 1933): 3.

28. Powdermaker, *Hollywood, the Dream Factory*, 210.

29. King, "Articulating Stardom," 47.

30. See "Unsung Specialists of the Screen Who Starve Plenty Between Calls," *Variety* (11 April 1933): 2.

31. Powdermaker, *Hollywood, the Dream Factory*, 254.

32. Ibid.

33. Raymond Williams, *Marxism and Literature* (New York: Oxford University Press, 1977): 195.

34. Ibid., 197.

35. Richard Johnson, "What Is Cultural Studies Anyway?" *Social Text* 6.1 (1987): 69.

36. Ibid.

37. Ibid., 45.

38. See "How the Cinematographer Works and Some of His Difficulties," *Motion Picture World* (8 June 1907): 212; also see "Is the Moving Picture to Be the Play of the Future?" *New York Times* (20 August 1911), as quoted in Richard deCordova, "The Emergence of the Star System in America," *Wide Angle* 6.4 (1985): 6.

39. DeCordova, "The Emergence of the Star System," 6.

40. Ibid., 10.

41. Alfred Harding, *The Revolt of the Actors* (New York: William Morrow, 1929): 286.

42. Louis B. Perry and Richard S. Perry, *A History of the Los Angeles Labor Movement, 1911-1941* (Berkeley: University of California Press, 1963): 337.

43. For a more complete analysis of Equity's role in Hollywood, see Danae A. Clark, "Actors' Labor and the Politics of Subjectivity: Hollywood in the 1930s" (Ph.D. diss., University of Iowa, 1989).

44. Murray Ross, *Stars and Strikes* (1941; reprinted, New York: AMS Press, 1967): 24.

45. Ibid., 7.

46. Harding, *The Revolt*, 286.

47. Ibid., 287.

48. Ross, *Stars and Strikes*, 24.

49. Harding, *The Revolt*, 356.

50. Ross, *Stars and Strikes*, 25.

51. Ibid., 26. Also see Harding, *The Revolt*, 533-34.

52. Harding, *The Revolt*, 535.

53. As quoted in ibid., 536.

54. As quoted in Leonard Mosley, *Zanuck* (New York: McGraw-Hill, 1984): 109.

55. Perry and Perry, *History of the Los Angeles Labor Movement*, 339.

56. Ibid., 320.

57. Harding, *The Revolt*, 536.

58. Ibid., 539.

59. Walker, *Stardom*, 212.

60. Ibid., 214.

61. Ibid., 218.

62. Harding, *The Revolt*, 540.

63. Ross, *Stars and Strikes*, 33.

64. Somerset Logan, "Revolt in Hollywood," *Nation* (17 July 1929): 62.

65. Perry and Perry, *History of the Los Angeles Labor Movement*, 339.

66. Logan, "Revolt in Hollywood," 62. "In these contracts there is no stipulation as to the length of the working day, or the length of the entire engagement. . . . Actors are quite frequently paid nothing for rehearsals. There are instances of players being required to work from sixty to eighty hours a week. When on location, any hours, from eight to twenty, have constituted a work day—sometimes with an additional bonus, sometimes not."

67. Ibid. As quoted in Logan, "Revolt in Hollywood," 62.

68. Ibid.

69. As quoted in "Unionism in Filmland," *Nation* (28 August 1929): 211.

70. Ross, *Stars and Strikes*, 31.

71. See Perry and Perry, *History of the Los Angeles Labor Movement*, 341.

72. Harding, *The Revolt*, 542-43.

3 / The Politics of (Self-) Representation

1. See Eric Lichten, "Crisis and Social Theory," *Class, Power and Austerity* (South Hadley, Mass.: Bergin & Garvey, 1986): 26-40.

2. Stuart Hall et al., *Policing the Crisis: Mugging, the State and Law and Order* (London: Macmillan, 1978): 217.

3. Lichten, "Crisis and Social Theory," 35.

4. Thomas Schatz, " 'A Triumph of Bitchery': Warner Bros., Bette Davis and *Jezebel*," *Wide Angle* 10.1 (1988): 16.

5. Hortense Powdermaker, *Hollywood, the Dream Factory* (Boston: Little, Brown, 1950): 327.

6. Ibid.

7. Frank Woods, "History of Producer-Talent Relations in the Academy," *Screen Guilds' Magazine* (November 1935): 4.

8. Ibid.

9. Ibid.

10. Ibid.

11. Louis B. Perry and Richard S. Perry, *A History of the Los Angeles Labor Movement, 1911-1941* (Berkeley: University of California Press, 1963): 342. Signed in February 1930 and renewed the following year for four years, the contract "provided for compulsory arbitration by the AMPAS conciliation committee of all disputes arising under the contract, a guarantee by the actors not to strike for the period of the agreement, and an eight-hour day with overtime for day players earning more than $15."

12. Murray Ross, *Stars and Strikes* (1941; reprint, New York: AMS Press, 1967): 44.

13. "Academy Close to a Break with Producers," *Equity* (May 1933): 9-10.

14. Ross, *Stars and Strikes*, 43.

15. Ibid., 89.

16. "Cagney Takes WB Cut; Chatterton Gives Free Film," *Variety* (14 March 1933): 5.

17. See Perry and Perry, *History of the Los Angeles Labor Movement*, 348. See also Woods, "History of Producer-Talent Relations," 4, 26.

18. "Industry's Reasons for Sweeping Cuts," *Variety* (14 March 1933): 5.

19. "No Production Shutdown," *Variety* (14 March 1933): 25.

20. Ibid.

21. "Films' Payrolls Drop Off 67%," *Variety* (14 March 1933): 7; Ross, *Stars and Strikes*, 45.

22. "Academy Close to a Break with Producers," 10.

23. "Films' Payrolls Drop Off 67%," 7.

24. "If Stars Walk, OK, Say Execs," *Variety* (7 March 1933): 3.

25. Lichten, "Crisis and Social Theory," 2.

26. Quoted in "No Production Shutdown," 25.

27. "Industry Leaders Wax Impatient with Personnel's Captiousness on Cuts; Most Being Overpaid?" *Variety* (28 March 1933): 5.

28. "New Players' Union Move," *Variety* (28 March 1933): 1.

29. "High-Priced Stars, Ritzy No More, Meet Unionizing Idea Half Way," *Variety* (4 April 1933): 3.

30. "Seek Academy, WB Accord on Cuts," *Variety* (18 April 1933): 5. To his credit, producer Darryl Zanuck resigned from Warner Bros. when he learned of the decision.

31. "100% Opposish to Producers' Plan," *Variety* (25 April 1933): 38.

32. "Industry Leaders Wax Impatient," 5.

33. "Academy Amending Rules, Retaining Producers in Fight with Radicals," *Variety* (2 May 1933): 2.

34. "Producers' Walkout on Academy Anticipated; Would Admit Agents to Solidify all H'Wood Branches," *Variety* (18 April 1933): 5.

35. "100% Opposish to Producers' Plan," 38; "Academy Amending Rules," 2.

36. "Academy Gains Membership Slowly Despite Bars Down," *Variety* (16 May 1933): 4.

37. "Academy Close to a Break with Producers," 10.

38. "Who Serves Who . . . and Who Pays?" *Screen Guilds' Magazine* (October 1935): 7, 20.

39. *Academy Bulletin* (29 July 1933): 1, as quoted in Ross, *Stars and Strikes*, 97.

40. "Actors' Branch Mulls Code of Ethics," *Variety* (11 July 1933): 4.

41. "Anti-Academy Ex-Stage Actors Forming Closed Shop Coast Union Similar to Equity Setup," *Variety* (25 July 1933): 5. See also "The Guild's Heritage," Screen Actors Guild pamphlet, n.d.

42. David F. Prindle, *The Politics of Glamour: Ideology and Democracy in the Screen Actors Union* (Madison: University of Wisconsin Press, 1988): 22.

43. *Variety* (25 July 1933): 5, 29. See also "Who Serves Who . . . and Who Pays?" 7.

44. "Only One Code Talked Over among Film Show People all through Past Week," *Variety* (27 June 1933): 4.

45. "Looks Like a Hays Code," *Variety* (27 June 1933): 5.

46. "All Branches of Industry Will Get a Chance at Code Framing," *Variety* (25 July 1933): 7.

47. "Not the Code for Motion Picture Actors," *Equity* (September 1933): 3.

48. "Coast Bums Not 'Extras,' " *Variety* (15 August 1933): 3.

49. "Academy Refused Data on Extras," *Variety* (22 August 1933): 2.

50. "Coast Bums Not 'Extras,' " 3.

51. Perry and Perry, *History of the Los Angeles Labor Movement*, 344.

52. Ross, *Stars and Strikes*, 104.

53. "Unionization of Extras by AFL Again," *Variety* (29 August 1933): 3.

54. Larry Ceplair and Steven Englund, *The Inquisition of Hollywood* (Berkeley: University of California Press, 1983): 28.

55. "Motion Picture Code Signed," *Equity* (December 1933): 6.

56. "Player Loans Still Mythical Despite Hays' Spirit of Co-op Talk," *Variety* (11 July 1933): 4. See also Ross, *Stars and Strikes*, 93.

57. See Ross, *Stars and Strikes*, 98.

58. "Academy Producers-Agents' Pact to Be Included in NRA Code, Opposed on Coast; Menjou's Agent Charges," *Variety* (29 August 1933): 7.

59. "Motion Picture Code Signed," 6.

60. "Who Serves Who . . . and Who Pays?" 20-21.

61. Francis L. Burt, "Advisory Boards Named to Hear Code Differences with Rosenblatt," *Motion Picture Herald* (2 September 1933): 26.

62. Ross, *Stars and Strikes*, 101. The star-raiding clause was modified to apply only to those employees who had been under contract for more than one year at a salary of $1,000 or more per week or employees who were under contract for three or more pictures at $15,000 or more per picture.

63. Ibid. See also Terry Ramsaye, "Problem of Labor Leads Debate at Capitol's Hearing on Code," *Motion Picture Herald* (16 September 1933): 11.

64. "Groups Strengthen Forces for Public Hearing on Code Tuesday," *Motion Picture Herald* (9 September 1933): 9.

65. See Ross, *Stars and Strikes*, 102-3.

66. Ramsaye, "Problem of Labor," 11.

67. See "Equity Mobilizes for Picture Hearing," *Equity* (September 1933): 5.

68. Ramsaye, "Problem of Labor," 11.

69. "Actors Defy Academy Groups in Battling Proposed Code Set-up," *Variety* (12 September 1933): 7. See also Ramsaye, "Problem of Labor," 11.

70. "Actors Defy Academy Groups," 7.

71. "Motion Picture Code Signed," 6.

72. Ramsaye, "Problem of Labor," 11.

73. "New Film-Actor Org.; Group Resigns from Academy," *Variety* (3 October 1933): 7. See also Terry Ramsaye, "Rosenblatt's Code Draft Long on Wage, Short on Disputed Points," *Motion Picture Herald* (7 October 1933): 10.

74. Ross, *Stars and Strikes*, 105.

75. "Actors' Secession Splits Academy; New Guild Members Nearing 1,000," *Motion Picture Herald* (14 October 1933): 10.

76. "The Beginning," *Screen Guilds' Magazine* (September 1935): 15.

77. "100% Screen Actors' Guild," *Variety* (10 October 1933): 7.

78. "The Wire to President Roosevelt and the Executive Order," *Screen Player* (June 1934): 4.

79. Ibid.

80. Terry Ramsaye, "Roosevelt Steps in while Industry Tries Again to Write Code," *Motion Picture Herald* (23 September 1933): 9.

81. "Screen Pay Curb Seen," *New York Times* (12 October 1933): 2:2.

82. Terry Ramsaye, "Rosenblatt Revises Code, but Roosevelt Asks about Those Salaries," *Motion Picture Herald* (14 October 1933): 16; "NRA to Study High Pay of Child Movie Stars," *New York Times* (12 October 1933): 2:2.

83. "Third Code Draft Issued, But Final Windup Is Still a Guess," *Motion Picture Herald* (21 October 1933): 10.

84. "Actors Threaten Strike in Movies," *New York Times* (13 October 1933): 24:6.

85. Ibid.

86. "Hollywood Salary Jams Obstacle to a Code Written by the Industry as New Complexities Cloud Outlook," *Variety* (26 September 1933): 7.

87. "Code Ducks Big Film Pay," *Variety* (10 October 1933): 5.

88. "Third Code Draft Issued," 10.

89. "Film Code Goes to White House; Salary Inquiry Is Made General," *Motion Picture Herald* (28 October 1933): 16.

90. "Third Code Draft Issued," 32.

91. "Code Is at 'Little White House' with Industry Interest Lagging," *Motion Picture Herald* (25 November 1933): 20.

92. "Text of Eddie Cantor's Speech at Annual Meeting," *Screen Player* (May 1934): 9.

93. Ibid.

94. See, for example, Philip Beck, "Historicism and Historism in Recent Film Historiography," *Journal of Film and Video* 37 (1985): 5-19; Janet Staiger and Douglas Gomery, "History of World Cinema: Models for Economic Analysis," *Film Reader* 4 (1980): 35-44.

95. Teresa de Lauretis, "Desire in Narrative," in *Alice Doesn't* (Bloomington: Indiana University Press, 1984): 106.

96. Ross, *Stars and Strikes*, 106.

97. P. W. Wilson, "Motion Pictures Move into the New Deal," *Literary Digest* (6 January 1934): 9.

98. Ramsaye, "Roosevelt Steps In," 1.

99. "Hollywood—Actors Tricked," *Nation* (16 January 1935): 59.

100. "The Guild's Heritage," pamphlet, Screen Actors Guild, n.d. See also the Producer-Screen Actors Guild Basic Minimum Contract of 1937.

101. "Screen Actors . . . Equity Contract," *Screen Guilds' Magazine* (December 1934): 20. See also "Guild: Movie Actors Win Long Fight and Unite with Equity," *Newsweek* (26 January 1935): 26.

102. Ross, *Stars and Strikes*, 163.

103. Eddie Cantor, "What the Guild Stands For," *Screen Player* (March 1934): 2.

104. Prindle, *The Politics of Glamour*, 14.

105. "Hollywood Clings to Huge Salaries," *New York Times* (16 October 1933): 15:7.

106. Figures from an NRA study as listed in Ross, *Stars and Strikes*, 108. See also "Hollywood—Actors Tricked," 59.

107. See Ross, *Stars and Strikes*, 163–72.

108. Ibid., 172.

109. *Screen Guilds' Magazine* (June 1937): 6.

110. Ceplair and Englund, *Inquisition of Hollywood*, 18.

111. Perry and Perry, *History of the Los Angeles Labor Movement*, 353; Ross, *Stars and Strikes*, 44.

112. Ross, *Stars and Strikes*, 173.

113. Cantor, "What the Guild Stands For," 2.

4 / Discourses of Entertainment

1. Report of the President to the Motion Picture Producers and Distributors of America, Inc., 27 March 1933.

2. Production Code Administration Annual Report of 1931.

3. Report of the President to the Motion Picture Producers and Distributors of America, Inc., 11 April 1932.

4. By-laws of the Motion Picture Producers and Distributors of America, Inc., Article 1, Section 3.

5. Will H. Hays, *The Memoirs of Will H. Hays* (Garden City, N.Y.: Doubleday, 1955): 447.

6. Ibid., 444.

7. Ibid., 445.

8. Raymond Moley, *The Hays Office* (Indianapolis: Bobbs-Merrill, 1945): 8.

9. Hays, *Memoirs*, 448.

10. Reaffirmation of the Code to Govern the Making of Talking, Synchronized and Silent Motion Pictures, 7 March 1933.

11. Ibid.

12. Hays, *Memoirs*, 448.

13. "Paradoxical Moral Code," *Variety* (29 August 1933): 7.

14. Ibid.

15. *Code of Fair Competition for the Motion Picture Industry* (Washington, D.C.: Government Printing Office, 1933): 255.

16. Daniel Bertrand, *Work Materials No. 34: The Motion Picture Industry* (Washington, D.C.: Government Printing Office, February 1936): 44.

17. Terry Ramsaye, "Problem of Labor Leads Debate at Capitol's Hearing on Code," *Motion Picture Herald* (16 September 1933): 19.

18. Bertrand, *Work Materials*, 44.

19. Ramsaye, "Problem of Labor," 19.

20. *Code of Fair Competition*, 249. Also see Bertrand, *Work Materials*, 83–87.

21. Moley, *Hays Office*, 52–53.

22. Ibid., 59.

23. Alexander Walker, *Stardom* (New York: Stein & Day, 1970): 196.

24. Congressional Record, 62:9657 (29 June 1922), as quoted in Moley, *Hays Office*, 27.

25. Nick Roddick, *A New Deal in Entertainment* (London: British Film Institute, 1983): 66.

26. Will H. Hays Collected Papers, Indiana State Library, Indianapolis, Indiana.

27. "Hollywood's Gab Geyser," *Variety* (3 October 1933): 2.

28. "Code Ducks Big Film Pay," *Variety* (10 October 1933): 5.

29. Jay B. Chapman, "Figuring the Stars' Salaries," *Screen Book* (November 1933): 11.

30. Ibid.

31. Ibid., 12.

32. Ibid.

33. Ibid., 49.

34. Jan Vantol, "How Stars Spend Their Fortunes," *Screen Book* (1933), as found in *Hollywood and the Great Fan Magazines*, ed. Martin Levin (New York: Arbor House, 1970): 12-13, 172.

35. Delight Evans, "Watch Your Step, Ann Dvorak!" *Screenland* (February 1933): 14.

36. Ibid., 15.

37. Ibid.

38. "Naughty Players Must Behave or Else," *Variety* (27 June 1933): 2.

39. "Doghouse for Baddies," *Variety* (5 September 1933): 3.

40. Donna Carla, "Behind the Scenes," *Screen Book* (February 1933): 42.

41. Elizabeth Wilson, "Watching the Stars at Work," *Silver Screen* (November 1933): 10.

42. Jack Jamison, "The Fascinating Mannerisms of the Stars," *Silver Screen* (March 1933): 64.

43. J. Eugene Chrisman, "Out of Tragedy to Happiness," *Screen Book* (November 1933): 27, 51.

44. Marquis Busby, "The Price They Pay for Fame," *Silver Screen* as found in *Hollywood and the Great Fan Magazines*, 94-96.

45. Chrisman, "Out of Tragedy," 27.

46. Grace Mack, "I've Learned Tolerance," *Screen Book* (December 1933): 35, 64-65.

47. Ben Maddox, "Gable! The Movies Saved Him!" *Screenland* (August 1933): 18-19, 74.

48. Ann Harding, "Thanks for the Buggy Rides," *Screen Player* (March 1934): 3.

49. As quoted in Elizabeth Yeaman, "The Other Side of the Moon," *Screen Guilds' Magazine* (November 1934): 8.

50. "The Call Board," *Screen Guilds' Magazine* (September 1934): 20.

51. Ibid.

52. Ibid.

53. Ann Harding, "Unique and Extraordinary," *Screen Guilds' Magazine* (September 1934): 3.

54. The methods of studio control can best be defined as "strategies," while those of actors' resistance are best defined as "tactics." See Michel de Certeau, *The Practice of Everyday Life* (Berkeley: University of California Press, 1988): 35-37.

55. "Cowan Lashes Creative Talent for Racketeering Pic Problems; Urges Own Wash Tub for Linen," *Variety* (31 October 1933): 3.

56. Ibid.

57. Quoted in "Actors Threaten Strike in Movies," *New York Times* (13 October 1933): 24:6.

58. "Cowan Lashes Creative Talent," 3.

59. *Historical Statistics of the U.S.*, Part I (Washington, D.C.: Government Printing Office, 1975): 179.

60. See, for example, "Relief Fund Growing," *Los Angeles Times* (22 February 1933): 5; and "Coast Film Relief Funds Carrying On," *Variety* (7 March 1933): 3

61. "Ether Pulling Curtain from Actors' Charity," *Variety* (14 February 1933): 2.

62. "Labors of Love," *Variety* (11 April 1933): 6.

5 / Labor and Film Narrative

1. Richard Johnson, "What Is Cultural Studies Anyway?" *Social Text* 6.1 (1987): 62.

2. Tom Gunning, "Film History and Film Analysis: The Individual Film in the Course of Time," *Wide Angle* 12.3 (1990): 6.

3. Ibid., 14.

4. Karl Marx, *Capital*, vol. 1, trans. Ben Fowkes (New York: Vintage, 1977): 130.

5. Martyn J. Lee, *Consumer Culture Reborn* (London: Routledge, 1993): 119.

6. Tony Bennett and Janet Woollacott, *Bond and Beyond* (London: Macmillan, 1987): 189.

7. Stuart Ewen, *Captains of Consciousness* (New York: McGraw-Hill, 1976): 100.

8. "Studio Prod. Is 75% Normal," *Variety* (1 August 1933): 5.

9. "Want to Be First for Roosevelt," *Variety* (25 June 1933): 7.

10. This Warner Bros. advertisement appeared in *Variety* (21 March 1933): 15.

11. This MGM advertisement appeared in *Variety* (3 October 1933): 28.

12. "Strike Unions Squawk to President," *Variety* (1 August 1933): 5.

13. "Unpatriotic to Pay for Meals, Overtime," *Variety* (8 August 1933): 1. See also, "Up for Extras," *Variety* (2 May 1933): 2; Eric L. Ergenbright, "All Over the Hollywood Lot," *Screen Book* (December 1933): 28.

14. "Harry Warner as NIRA Propagandist," *Variety* (1 August 1933): 5.

15. "Warner's Semi-official 'New Deal' NRA Short," *Variety* (1 August 1933): 1.

16. "64,000 Showings per NRA Short Scheduled," *Variety* (29 August 1933): 3.

17. "Eight NRA Films Made; One Release a Week," *Motion Picture Herald* (16 September 1933): 22.

18. "250,000 March under the Blue Eagle in the City's Greatest Demonstration," *New York Times* (14 September 1933): 3.

19. Motion Picture Producers and Distributors Association Annual Report (11 April 1932): 24.

20. "Eight NRA Films Made," 22.

21. "Film Shorts for NRA Shown," *Variety* (5 September 1933): 4.

22. "Femme Film Workers Protest to Legislature against 8-Hr. Day," *Variety* (2 May 1933): 2.

23. "No Blue Eagle Blues," *Variety* (29 August 1933): 2.

24. "Film Shorts for NRA Shown," 4. *The New Deal*, the original title proposed by Warner Bros., did not become part of this NRA series.

25. Andrew Bergman, *We're in the Money* (New York: Harper & Row, 1971): 64.

26. For example, see John Belton, "The Backstage Musical," *Movie* 24 (1977): 36–43; Bergman, *We're in the Money*; Rocco Fumento, "Those Berkeley and Astaire-Rogers Depression Musicals: Two Different Worlds," *American Classic Screen* 5.4 (1981): 15–18; Mark Roth, "Some Warners Musicals and the Spirit of the New Deal," in *Genre: The Musical*, ed. Rick Altman (London: Routledge & Kegan Paul, 1981): 41–56. For a feminist analysis, see Paula Rabinowitz, "Commodity Fetishism: Women in *Gold Diggers of 1933*," *Film Reader* 5 (1982): 141–49.

27. Roth, "Some Warners Musicals," 45.

28. "WB Sets High Figure for '42nd St.' if RKO Wants Special Train Angle," *Variety* (14 February 1933): 7.

29. Ibid.

30. Charles Eckert, "The Carole Lombard in Macy's Window," *Quarterly Review of Film Studies* 3.1 (1978): 3.

31. "WB Sets High Figure," 7.

32. Rocco Fumento, ed., "Introduction: From Bastards and Bitches to Heros and Heroines," in *42nd Street*. Wisconsin/Warner Bros. Screenplay Series (Madison: University of Wisconsin Press, 1980): 35.

33. Warner Bros. Archives, press kit, University of Southern California, Los Angeles.

34. "Ruby Keeler Would Dance, Not Hand Out S. A. in WB's 'Footlight Parade,' " *Variety* (25 July 1933): 2.

35. Warner Bros. Archives, memo from R. J. Obringer to Mr. Sully (18 February 1933).

36. Contract of Virginia Dabney (18 February 1933).

37. Warner Bros. Archives, press kit.

38. Contract of Ginger Rogers (15 September 1932).

39. "The Guild's Heritage," Screen Actors Guild pamphlet, n.d.

40. "Chorus Girls Toil 86 Hours for $25 Week, Hearing Told," *Motion Picture Herald* (16 September 1933): 21. Also see "Are Extras People?" *Screen Guilds' Magazine* (November 1935): 25.

41. "Chorus Girls Toil 86 Hours," 21.

42. Roth, "Some Warners Musicals," 45.

43. Warner Bros. Archives, letter from Guy B. Kibbee to Warner Bros. Pictures, Inc., n.d.

44. Warner Bros. Archives, letter from Warner Bros. Pictures, Inc., to Guy Kibbee (15 April 1933). Although the two letters obviously refer to the same contract, Kibbee dates the contract March 16, 1931 and Warners dates it May 16, 1931.

45. Thomas Schatz, " 'A Triumph of Bitchery': Warner Bros., Bette Davis and *Jezebel*," *Wide Angle* 10.1 (1988): 18.

46. Warner Bros. Archives, press kit.

47. Warner Bros. Archives, letter from Frank Joyce to Jack Warner (11 February 1932).

48. Warner Bros. Archives, affidavit re: Bebe Daniels (21 April 1932).

49. Warner Bros. Archives, letter from Warner Bros. Pictures, Inc., to Fox Film Corporation; Metro-Goldwyn-Mayer Corp.; RKO Studios, Inc.; Columbia Pictures, Corp.; United Artists Studios; Universal Pictures Corporation; Samuel Goldwyn Inc. Ltd.; Paramount Productions, Inc.; Tiffany Productions of Calif. Ind. Ltd.; Mr. Fred Beetson, Assn. of Motion Picture Producers (30 October 1933).

50. See "Zanuck's Raiding Mess," *Variety* (13 June 1933): 5, 43; "Harry Warner Says Battle Is Now On," *Variety* (27 June 1933): 5.

51. Jane Feuer, "The Self-Reflective Musical and the Myth of Entertainment," in *Genre: The Musical*, ed. Rick Altman, 159, 162.

52. Alvin H. Marill, *Katharine Hepburn* (New York: Galahad, 1973): 26.

53. "Dos and Don'ts for Equity Members," *Equity* (June 1933): 19.

54. "Stars Have Responsibilities as well as Privileges," *Equity* (May 1933): 18.

55. Lew Garvey, "Fleeting Fame," *Screen Book* (November 1933): 30.

56. Tichi Wilkerson and Marcia Borie, "Extra Girl Gets Her First Close-up," *The Hollywood Reporter: The Golden Years* (New York: Coward-McCann, 1984): 61.

57. "Unemployment Fear from H'Wood-at-Fair," *Variety* (23 May 1933): 6.

58. "Peeved Beaut Contest Losers Storm H'wood with Gate-Crashing Wiles," *Variety* (3 October 1933): 3.

59. "Hungry Need for New Film Faces, Every Studio on Talent Hunt," *Variety* (10 October 1933): 3; "Talent Famine, and No Kiddin'," *Variety* (5 December 1933): 1.

60. "Studios Revive School System," *Variety* (10 October 1933): 3.

61. "Actors' Chances Put at 10 to 1," *Variety* (9 May 1933): 7.

62. "Coast Bums Not 'Extras,' " *Variety* (15 August 1933): 3.

63. Ibid.

64. "Oldtimers Topping Studios' 1st Call List," *Variety* (1 August 1933): 1.

65. "Coast Bums Not 'Extras,' " 3.

66. "Sign over 200 Pic Players, Inflation?" *Variety* (19 September 1933): 2.

67. Jay B. Chapman, "Hollywood's New Beauty Standards," *Screen Book* (August 1933): 11-13.

68. "Looks, Not Talent, Still Gets Tests," *Variety* (26 September 1933): 2.

69. "Footlite Talent Tops Active List," *Variety* (19 September 1933): 1.

70. Constance McCormick Collection, University of Southern California, Los Angeles.

71. "Sweat Shop Charge of Penny-Ante Pay Takes Extras' Plight to Code," *Variety* (12 September 1933): 7.

72. Ibid.

73. "The Letter Box," *Screen Player* (June 1934): 20-21.

74. "A Jeer," *Screen Player* (June 1934): 13.

75. "Cheers and Jeers," *Screen Player* (March 1934): 7.

76. "Are Extras People?" *Screen Guilds' Magazine* (November 1935): 3.

77. Ibid.

78. "Shop Talk," *Screen Guilds' Magazine* (October 1936): 3.

79. This information is taken from a *Morning Glory* advertisement, *Variety* (22 August 1933): 32.

6 / The Terrain of Actors' Labor

1. Lawrence Grossberg, *We Gotta Get Out of This Place* (New York: Routledge, 1992): 20-21.

2. Ibid., 54.

3. Ibid., 54-55.

4. Barry King, "The Star and the Commodity: Notes Towards a Performance Theory of Stardom," *Cultural Studies* 1.2 (1987): 146.

5. Richard Dyer, *Heavenly Bodies* (New York: St. Martin's Press, 1986): 2,

6. Richard Dyer, *Stars* (London: British Film Institute, 1979): 37.

7. Dyer, *Heavenly Bodies*, 141-56.

8. John Fiske, *Reading the Popular* (Boston: Unwin Hyman, 1989): 98.

9. Cathy Schwichtenberg, ed., *The Madonna Connection* (Boulder, Colo.: Westview Press, 1992).

10. King, "The Star and the Commodity," 147-48.

11. Mike Budd, Robert M. Entman, and Clay Steinman, "The Affirmative Character of U.S. Cultural Studies," *Critical Studies in Mass Communication* 7.2 (1990): 170.

12. A few critical theorists have examined the way in which audiences for commercial television labor as consumers on behalf of the advertising industry. For a summary of this position, see Oscar Gandy, "Tracking the Audience," in *Questioning the Media*, ed. John Downing, Ali Mohammadi, and Annabelle Sreberny-Mohammadi (Newbury Park, N.J.: Sage, 1990): 170-71.

13. King, "The Star and the Commodity," 158.

14. Dyer, *Heavenly Bodies*, 6.

15. King, "The Star and the Commodity," 149.

16. Ronald Brownstein, *The Power and the Glitter* (New York: Pantheon, 1990): 8.

17. Laurie Winer, "Cause Celeb," *Mirabella* (February 1992): 86.

18. Brownstein, *Power and the Glitter*, 11.

19. Letters to the Editor, *Cityflickers* (October 1992): 2.

20. "Outlook Dim for Stand-Ins," *Variety* (10 October 1933): 3.

21. Monte Williams, "Industry Calls Body Doubles' Gripes a Naked Grab for Power," *Pittsburgh Post-Gazette* (7 September 1993): B3.

22. Ibid.

23. Ibid.

24. Ibid.

25. Dyer, *Heavenly Bodies*, 6.

Selected Bibliography

Affron, Charles. *Star Acting*. New York: Dutton, 1977.

Bazin, André. *What Is Cinema?* Trans. Hugh Gray. Berkeley: University of California Press, 1967.

Belton, John. "The Backstage Musical." *Movie* 5 (1977): 36–43.

Bennett, Tony. "Theories of the Media, Theories of Society." In *Culture, Society and the Media*. Ed. Michael Gurevitch, Tony Bennett, James Curran, and Janet Woollacott. London: Methuen (1982): 30–55.

Bennett, Tony, and Janet Woollacott. *Bond and Beyond*. New York: Methuen, 1987.

Bergman, Andrew. *We're in the Money*. New York: Harper & Row, 1971.

Bertrand, Daniel. *Work Materials No. 34: The Motion Picture Industry*. Washington, D.C.: Government Printing Office, February, 1936.

Bordwell, David, Janet Staiger, and Kristin Thompson. *The Classical Hollywood Cinema: Film Style and Mode of Production to 1960*. London: Routledge & Kegan Paul, 1985.

Britton, Andrew. *Katharine Hepburn*. Newcastle upon Tyne: Tyneside Press, 1984.

Brownstein, Ronald. *The Power and the Glitter*. New York: Pantheon, 1990.

Budd, Mike, Robert M. Entman, and Clay Steinman. "The Affirmative Character of U.S. Cultural Studies." *Critical Studies in Mass Communication* 7.2 (1990): 169–84.

Burawoy, Michael. *The Politics of Production*. London: Verso, 1985.

Butler, Jeremy G., ed. *Star Texts*. Detroit: Wayne State University Press, 1991.

Cantor, Eddie. "What the Guild Stands For." *Screen Player* (March 1934): 2.

Ceplair, Larry, and Steven Englund. *The Inquisition of Hollywood*. Berkeley: University of California Press, 1983.

Clark, Danae A. "Actors' Labor and the Politics of Subjectivity: Hollywood in the 1930s." Ph.D. Diss., University of Iowa, 1989.

Code of Fair Competition for the Motion Picture Industry. Washington, D.C.: Government Printing Office, 1933.

Cook, Pam. "Star Signs." *Screen* 20.3/4 (1979/80): 80–88.

deCordova, Richard. "The Emergence of the Star System in America." *Wide Angle* 6.4 (1985): 4–13.

———. *Picture Personalities: The Emergence of the Star System in America*. Urbana: University of Illinois Press, 1990.

Doty, Alexander. "The Cabinet of Lucy Ricardo: Lucille Ball's Star Image." *Cinema Journal* 29.4 (1990): 3–22.

Dyer, Richard. *Heavenly Bodies*. New York: St. Martin's Press, 1986.

———. *Stars*. London: British Film Institute, 1979.

Eckert, Charles. "Shirley Temple and the House of Rockefeller." *Jump Cut* 2 (1974): 1, 17–20.

Ellis, John. "Star/Industry/Image." In *Star Signs*. Ed. Christine Gledhill. London: British Film Institute (1982): 1–12.

Ewen, Stuart. *Captains of Consciousness*. New York: McGraw-Hill, 1976.

Feuer, Jane. "The Self-Reflective Musical and the Myth of Entertainment." In *Genre: The Musical*. Ed. Rick Altman. London: Routledge & Kegan Paul (1981): 159-74.

Fiske, John. *Reading the Popular*. Boston: Unwin Hyman, 1989.

Friedberg, Anne. "Identification and the Star: A Refusal of Difference." In *Star Signs*. Ed. Christine Gledhill. London: British Film Institute (1982): 47-54.

Fumento, Rocco. "Those Berkeley and Astaire-Rogers Depression Musicals: Two Different Worlds." *American Classic Screen* 5.4 (1981): 15-18.

――――. "Introduction: From Bastards and Bitches to Heros and Heroines." In *42nd Street*. Wisconsin/Warner Bros. Screenplay Series. Madison: University of Wisconsin Press, (1980): 9-38.

Gaines, Jane. *Contested Culture: The Image, the Voice, and the Law*. Chapel Hill: University of North Carolina Press, 1991.

――――. "In the Service of Ideology: How Betty Grable's Legs Won the War." *Film Reader* 5 (1982): 47-59.

Garnham, Nicholas. "Subjectivity, Ideology, Class and Historical Materialism." *Screen* 20.1 (1979): 121-34.

Gledhill, Christine, ed. *Star Signs*. London: British Film Institute, 1982.

Grossberg, Lawrence. *We Gotta Get Out of This Place*. New York: Routledge, 1992.

Gunning, Tom. "Film History and Film Analysis: The Individual Film in the Course of Time." *Wide Angle* 12.3 (1990): 4-19.

Hall, Stuart. "Culture, the Media and the 'Ideological Effect.' " In *Mass Communication and Society*. Ed. James Curran, Michael Gurevitch, and Janet Woollacott. London: Sage (1979): 336-42.

――――. "The Rediscovery of 'Ideology': Return of the Repressed in Media Studies." In *Culture, Society and the Media*. Ed. Michael Gurevitch, Tony Bennett, James Curran, and Janet Woollacott. London: Methuen (1982): 56-90.

Hall, Stuart, et al. *Policing the Crisis: Mugging, the State and Law and Order*. London: Macmillan, 1978.

Harding, Alfred. *The Revolt of the Actors*. New York: William Morrow, 1929.

Hays, Will H. *The Memoirs of Will H. Hays*. Garden City, N.Y.: Doubleday, 1955.

Jacobowitz, Florence. "Joan Bennett: Images of Femininity in Conflict." *CineAction!* 7 (1986): 22-34.

Jayamanne, Laleen. "Modes of Performing: Bodies & Texts (some thoughts on fe/male performances)." *Australian Journal of Screen Theory* 9/10 (1981): 123-39.

Johnson, Richard. "What Is Cultural Studies Anyway?" *Social Text* 6.1 (1987): 38-80.

Kindem, Gorham. "Hollywood's Movie Star System: A Historical Overview." In *The American Movie Industry*. Ed. Gorham Kindem. Carbondale: Southern Illinois Press (1982): 79-93.

King, Barry. "Articulating Stardom." *Screen* 26.5 (1985): 27-50.

――――. "The Star and the Commodity: Notes Towards a Performance Theory of Stardom." *Cultural Studies* 1.2 (1987): 145-61.

――――. "Stardom as an Occupation." In *The Hollywood Film Industry*. Ed. Paul Kerr. London: Routledge (1986): 154-84.

Lee, Martyn J. *Consumer Culture Reborn*. London: Routledge, 1993.

Levin, Martin, ed. *Hollywood and the Great Fan Magazines*. New York: Arbor House, 1970.

Lichten, Eric. *Class, Power and Austerity*. South Hadley, Mass.: Bergin & Garvey, 1986.

Lippe, Richard. "Kim Novak: A Resistance to Definition." *CineAction!* 7 (1986): 4-21.

Lovell, Terry. *Pictures of Reality*. London: British Film Institute, 1983.

Maltby, Richard. " 'Baby Face' or How Joe Breen Made Barbara Stanwyck Atone for Causing the Wall Street Crash." *Screen* 27.2 (1986): 22-45.

Marill, Alvin H. *Katharine Hepburn*. New York: Galahad, 1973.

Marx, Karl. *Capital*. Volume I. Trans. Ben Fowkes. New York: Vintage, 1977.

————. *Wage-Labour and Capital* and *Value, Price and Profit*. New York: International Publishers, 1976.

Moley, Raymond. *The Hays Office*. Indianapolis: Bobbs-Merrill, 1945.

Morley, David. "Text Readers, Subjects." In *Culture, Media, Language*. Ed. Stuart Hall, Dorothy Hobson, Andrew Lowe, and Paul Willis. London: Hutchinson (1984): 163-73.

Mosco, Vincent. "Critical Research and the Role of Labor." *Journal of Communication* 33.3 (1983): 237-48.

Mouffe, Chantal. "Hegemony and New Political Subjects: Toward a New Concept of Democracy." In *Marxism and the Interpretation of Culture*. Ed. Cary Nelson and Lawrence Grossberg. Urbana: University of Illinois Press (1988): 89-101.

Munsterberg, Hugo. *The Photoplay: A Psychological Study*. New York: Appleton, 1916.

Naremore, James. *Acting in the Cinema*. Berkeley: University of California Press, 1988.

Nielson, Michael. "Toward a Worker's History of the U.S. Film Industry." In *The Critical Communications Review, Volume I: Labor, the Working Class, and the Media*. Ed. Vincent Mosco and Janet Wasko. Norwood, N.J.: Ablex (1983): 47-83.

Perry, Louis B., and Richard S. Perry. *A History of the Los Angeles Labor Movement, 1911-1941*. Berkeley: University of California Press, 1963.

Powdermaker, Hortense. *Hollywood, the Dream Factory*. Boston: Little, Brown, 1950.

Prindle, David. *The Politics of Glamour: Ideology and Democracy in the Screen Actors Guild*. Madison: University of Wisconsin Press, 1988.

Roddick, Nick. *A New Deal in Entertainment*. London: British Film Institute, 1983.

Ross, Murray. *Stars and Strikes*. 1941. Reprinted. New York: AMS Press, 1967.

Roth, Mark. "Some Warners Musicals and the Spirit of the New Deal." In *Genre: The Musical*. Ed. Rick Altman. London: Routledge & Kegan Paul (1981): 41-56.

Schatz, Thomas. " 'A Triumph of Bitchery': Warner Bros, Bette Davis and *Jezebel*." *Wide Angle* 10.1 (1988): 16-29.

Schwichtenberg, Cathy, ed. *The Madonna Connection*. Boulder, Colo.: Westview Press, 1992.

Staiger, Janet. "Seeing Stars." *Velvet Light Trap* 20 (1983): 10-14.

Turner, Graeme. *British Cultural Studies*. Boston: Unwin Hyman, 1990.

Walker, Alexander. *Stardom: The Hollywood Phenomenon*. New York: Stein & Day, 1970.

Watney, Simon. "Katharine Hepburn and the Cinema of Chastisement." *Screen* 26.5 (1985): 52-62.

Wilkerson, Tichi, and Marcia Borie. *The Hollywood Reporter: The Golden Years*. New York: Coward-McCann, 1984.

Williams, Raymond. *Marxism and Literature*. New York: Oxford University Press, 1977.

Woods, Frank. "History of Producer-Talent Relations in the Academy." *Screen Guilds' Magazine* (November 1935): 4, 26.

Yacowar, Maurice. "An Aesthetic Defense of the Star System in Films." *Quarterly Review of Film Studies* 4.1 (1979): 39-52.

Zucker, Carole, ed. *Making Visible the Invisible*. Metuchen, N.J.: Scarecrow Press, 1990.

Index

147

Danae Clark is currently associate professor of media studies in the Department of Communication at the University of Pittsburgh. Her work has appeared in *Camera Obscura*, the *Journal of Film and Video*, and other media journals. Her other research interests include gay and lesbian studies and consumer culture studies.